THE WISDOM OF THE BRAIN

THE WISDOM OF THE BRAIN

neuroscience for helping professionals

Camaron J. Thomas, Ph.D.
Edited by Franklin R. Amthor Ph.D.

Copyright © 2016 Camaron J. Thomas, Ph.D.
All rights reserved.

ISBN: 1537185608
ISBN 13: 9781537185606
Library of Congress Control Number: 2016914186
CreateSpace Independent Publishing Platform
North Charleston, South Carolina

Publisher's note

Every possible effort has been made to ensure that the information contained in this book is accurate at the time of going to press, and the publishers and author cannot accept responsibility for any errors or omissions, however caused. No responsibility for loss or damage occasioned to any person acting, or refraining from action, as a result of the material in this publication can be accepted by the editor, the publisher, or the author.

*For My Sister,
With Love*

Table of Contents

Publisher's note · v
Preface · xv

Section I	**An Introduction to Neuroscience** · · · · · · · · · · · · · · · 1	
Chapter 1	The Brain Manages the Environment · · · · · · · · · · · · · · 3	
Chapter 2	A Brief History · 11	
	The Conventional Model of the Brain · · · · · · · · · · · · · · 12	
	The History of the Larger Human Brain · · · · · · · · · · · · 14	
	Modern-Day Neuroscience · 17	
	Historical Know-How · 19	
Chapter 3	Hardwired Human Nature · 22	
	Hardwired Pathways and Instincts · · · · · · · · · · · · · · · 22	
	Hardwired Drives and Human Universals · · · · · · · · · · 24	
	A Tougher Nut to Crack · 25	
	The Brain's Three Special Properties · · · · · · · · · · · · · · 26	
	Hardwired or Not? · 28	
Section II	**How the Brain Works** · 31	
Chapter 4	Neural Pathways of Communication · · · · · · · · · · · · · · 33	
	What's All the Fuss about Neurons? · · · · · · · · · · · · · · 33	
	Neurons Form Pathways · 35	
	Our Subjective Inner State · 39	

Chapter 5	The Human Nervous System	42
	An Overview	43
	Pathways that Change	48
Chapter 6	The Organization of the Brain	51
	Regions of the Brain	53
	The Prefrontal Cortex in Greater Detail	60
Chapter 7	How the Brain Might Work	64
	The Computational Theory of Mind	65
	The Network Dynamics Model	68
Chapter 8	The Human Brain and What Makes Us Unique	72
	What Sets the Brain Apart	73
Section III	**How the Brain Thinks**	81
Chapter 9	Thinking as Representations	83
	Patterns First	83
	The Representational Approach to Thinking	85
	Representational Thinking and the Vision System	86
	Representational Thinking and Taking Action	87
Chapter 10	Thinking as a Cycle	91
	The Perception-Action Cycle Approach to Thinking	91
	A Subcortical Approach to Thinking	93
	Some Additional Thoughts on Thinking	95
Section IV	**The Brain and Individuality – Parenting, Emotion, and Past**	99
Chapter 11	Parenting and How the Brain Develops	101
	Evolution and Nurturance	102
	The Development of the Brain	103
Chapter 12	Parenting and the Brain	106
	Ready to Learn and Nurture	106
	Parenting Builds *Biology*	108
	Critical Windows and Conceptual Doors	110
	A Noteworthy Point	112

Chapter 13	How Parenting Builds the Brain · · · · · · · · · · · · · · · · · · · 113
	The Nature of Attachment · 114
	Attachment and Self-Control · 117
	Attachment and Self-Esteem · 118
	Attachment and Relationships · 119
	No Words to Question, No Capacity to Choose · · · · · · · 121
Chapter 14	The Brain and Emotion · 122
	What We Know about Emotions · 122
	More about Emotions · 125
	The Many Players in Emotion · 126
Chapter 15	The Many Faces of Emotion · 132
	Emotional Operating Systems · 132
	Emotion and the Cortex · 134
	Emotion and the Amygdala · 136
	Emotion and Decision-Making · 137
Chapter 16	Emotional Reactions · 142
	Conditioned Fear · 142
	The Threat Response · 145
	Stress as Re-action · 148
	An Emotional Analysis · 149
Chapter 17	This Too Shall Pass · 152
	The Story of "Me" · 153
	The Left Hemisphere Interpreter · · · · · · · · · · · · · · · · · · · 155
	A Filter of Past · 157

Section V	**The Brain and How We Accumulate Knowledge** · · · · · 161
Chapter 18	How the Brain Learns · 163
	It *Can* Be Taught · 163
	The Learning Brain · 165
	Learning at the Cellular Level · 168
	Entirely Plastic? · 169
Chapter 19	How the Brain Remembers · 171
	Types of Memory and the Brain Systems that Support Them · 173

		Working Memory – "The Blackboard of the Mind"	175
		Explicit and Implicit Memory	176
		When Memory Fails	180
Section VI		**Conscious and Unconscious Influences**	185
Chapter 20	The Scope of the Unconscious	187
		The Principles of the Brain	187
		The Work of the Unconscious	188
		Unconscious Influences	189
		The New Unconscious	193
		Habitual Behavior	195
		Take a *Conscious* Breath	197
Chapter 21	Making a Conscious Effort	201
		Consciousness and Intelligence	202
		Consciousness and Free Will	204
		Consciousness and Control	206
		Consciousness Explained Better[48]	207
Section VII		**A Brain that's Also Social**	211
Chapter 22	Social Cognition	213
		A Distinctly Social Orientation	213
		Social Neuroscience	215
Chapter 23	Social Beings, Social Mind	221
		The Origins of Human Sociality	221
		The Power of Culture	222
		Individuals Equipped With Selves	224
		Group Membership	225
		Self-Regulation	226
Chapter 24	Reading Other People	230
		Processing Faces	230
		Processing Social Categories	231
		Processing Social Pain	232
		Processing Language and Communication	233
		Processing…	234

Chapter 25	Theories about the Mind	237
	Theory of Mind and the Brain	238
	Developing a Theory of Mind	239
	How We Read a Person's Mind	242
	An Art Not a Science	245
Chapter 26	Theory of Mind in Action: Empathy	247
	A Problem of Definitions	247
	The Neural Bases of Empathy	249
	Emotional and Cognitive Empathy	250
	Thinking about Empathy	251
	In Sum	253
Chapter 27	Neuroscience and the Future of the Helping Professions	258
	The Five Summary Features and Where They Point	258
	In Closing	265
	Appendix A	267
	Author Biography	269
	References	271
	Index	317

Preface

Perhaps we should start with the question why: Why should neuroscience matter to helping professionals? Neuroscience is the study of the nervous system, including the brain. It offers a different and deeper look at human nature, and teaches us why people behave as they do. With the advent of *social* neuroscience, this field promises to play a prominent role in the future of social work, psychology, mediation, education, personal wellness, and all the other helping professions. To help navigate this future, The Wisdom of the Brain serves as a bridge...between the social and the neural sciences; what is now known in neuroscience and the practical applications which are soon to follow and will change how we do our jobs; and the technical aspects of the brain and the wisdom it has to offer.

Sometimes what we learn from neuroscience validates what we know. Sometimes it leads us down a completely new and untraveled path. On occasion, it reveals a startling insight about something we've always wondered about or thought we understood. Let's take an example. Suppose you work with habitual behavior – addiction, compulsive behavior, acting out, etc. You already know about cues, and the role of dopamine and rewards. By studying neuroscience, you will learn how habits are based on patterns and all thinking is grounded in patterns; that there is a part of the brain, called the basal ganglia, that is literally a habit-maker, and it works with other networks that motivate and guide behavior and monitor success or failure in obtaining a reward; that the release of dopamine not only sends the message "Do it again" and creates a craving for the reward, but dopamine neurons learn to gauge rewards levels, comparing anticipated rewards with actual ones and thereby direct future activity towards the most rewarding pursuits.[1]

You will also learn how unconscious influences and early childhood experiences set the stage for habitual behavior; how memory links like events together which may, in treatment, make the setting as important of a focus as the habit itself. You will understand how the series of steps in a habit unfold automatically, how learning depends on rehearsal, and that conscious awareness needs to be engaged – all of which make habits very hard to break. You'll learn that because of plasticity, the brain can change...but within reason; that some things are harder to change and most things never even reach consciousness. You'll come to appreciate that an estimated 45% of what we do every day is done "in the same environment and is repeated"[2] and ultimately, the routine gains the power to override intentions...that the brain seems to naturally revert to the path of least effort...and one option for overcoming a habit is to try to make self-control itself, an *unconscious* habit.[3]

The point is neuroscience provides a more in-depth examination of the situations, conditions, behaviors, thoughts, and problems that we, as helping professionals, face every day. Now perhaps, you've had a less than positive experience with neuroscience. You may have been on the receiving end of some "neuroscience lite" at a conference or seminar; maybe the material was dummied down, or the presenter avoided explaining the science, or the presentation was tedious or worse, irrelevant. <u>The Wisdom of the Brain</u> is specifically designed for professionals who help others:

- It targets the topics of interest to helping professionals: from parenting, to stress, to how we read other people;
- It accumulates in one text, all the latest and most relevant information so you don't have to hunt around for it; and,
- It offers just the right balance of science and insight that respects the intelligence of the reader but doesn't cause science-overload.

Neuroscience is still science though, so every effort has been taken to make this book as easy to read as possible. The book is written in compact chapters that can be read in a single sitting. It introduces the regions of the brain and uses their proper names, highlighting the terms in bold whenever necessary. On occasion, the proper abbreviations are also presented, especially for those areas of the brain that will likely come up in future conversations. To facilitate further study, additional books to read are recommended. And many chapters end with "A Closing Thought" – a discussion of some of the most pressing questions and quandaries facing neuroscience today. Questions include the relationship between the mind and the brain, how real reality might be, whether the study of neuroscience undermines the belief in God, and if we are nothing more than the sum of our neurons.

<u>The Wisdom of the Brain</u> is divided into seven sections. The first introduces the study of neuroscience, offers some interesting history, and examines what aspects of human nature may be hardwired. The second looks at how the brain works. It covers the various regions of the brain, the subjective inner milieu, neuroplasticity, why the prefrontal cortex is *not* the executive brain, and includes a chapter on how and why the brain is not like a computer. Section three considers how the brain thinks – through computations, linking inputs to representations, and moving from perception to action. The fourth section examines how we become individuals: how parenting shapes neural pathways, the role emotions play, and the development of a sense of self, a filter of past, and a left hemisphere interpreter who narrates our life story. Section five addresses how we learn and remember, while section six considers the breadth of unconscious influences that dominate our lives – including an area called "the new unconscious" – as well as the role of conscious awareness. Social cognition is covered in detail in the final section, including how we read other people and how empathy happens in the brain.

Most importantly, each chapter features a piece of wisdom of the brain especially for the benefit of helping professionals. The wisdom of the brain includes insights into why re-activity is our first line of defense;

how we can, at best, know the experience of another indirectly and in many respects, we barely know ourselves; and the three goals of the brain in every situation. Together, these pieces of wisdom -- while not prescriptive because neuroscience is still too young to derive specific recommendations -- point to possible new directions in the helping fields and changes that may be underway in what it means to help.

Of course, a book like this doesn't just happen; especially not to someone who works as a helping professional and is not a neuroscientist. Years ago, Dr. Frank Amthor, who wrote <u>Neuroscience for Dummies</u>, responded to an email of mine which he probably regrets doing. Dr. Amthor directs the Behavioral Neuroscience Graduate Program at the University of Alabama at Birmingham. Dr. Amthor was kind enough to vet the neuroscience sections of this book and guided my study of the field. He is working on some fascinating projects on deep brain stimulation and in other areas, and no matter what I say about this gentleman, it would be an understatement. I am forever and deeply grateful for his effort, patience, encouragement, and consistency. Dr. Cozolino's work on the neuroscience of human relationships was also very helpful. I am indebted to all of those who wrote wonderful books about the field: Dr. Eric Kandel, Drs. Sporns and Panksepp, Dr. Gazzaniga, Drs. Siegel and Lieberman, Dr. Porges, Dr. LeDoux – and I could go on. On equal footing, I have to thank the many people who have read and reread this book; my proof-reader and best friend, Ray Michaels, who never tired of being the keeper of the book; Ms. Ling Nah Su who read the text numerous times, and encouraged and cheered me on; and Ms. Sandy Nagel-Germain who introduced me to the idea of mind patterns. A special thank you needs to be extended to Carol A. Thomas, who saved me from a technology standpoint more often than I care to admit. And finally, thank you to my dear husband who has put up with everything that goes into writing a book about neuroscience. Thank you all.

This book has been written with you, the reader, in mind. It provides what you need to know about a field that promises to make us better -- better at doing one of the hardest jobs in the world, that of helping others.

Section I

An Introduction to Neuroscience

1

The Brain Manages the Environment

Neuroscience examines human nature from the inside out -- it examines how the biology, psychology, social nature, *and wisdom* of the brain shape human behavior. Studying neuroscience promises to change how we see ourselves; add depth to every interaction we have; and alter how we relate to others, what we expect of and from them. In short, it will make us better helping professionals.

Neuroscience is the study of the **nervous system** which commands movement. Plants don't have nervous systems. A violet can tilt its leaves towards the sun but it can't get up and walk away. Nervous systems allow organisms to move in a coordinated way, putting one foot in front of the other, galloping in stride, or flapping wings in flight. The human nervous system is comprised of the **central nervous system**: the brain, spinal cord, and the retina in the eye. The spinal cord connects the central nervous system to the **peripheral nervous system**, the neurons of which extend to the muscles, organs, etc. Humans also have an **autonomic nervous system** which handles internal matters such as breathing, digestion, heart rate, etc. The autonomic nervous system is "non-voluntary," meaning it does not require conscious control. Fight or flight originates in the sympathetic part of the autonomic nervous system; it is calmed and restored to balance through the parasympathetic part of this system.

The nervous system is "populated" by **neurons**. Neurons are special cells: they communicate with one another and translate input into information that is used by circuits of neurons throughout the body. Those circuits allow us to think, feel, and *move*. The human motor system provides us with a basic model of how the brain works. Let's say you're taking a walk:

> Walking along, you're probably using a sequence of movements you've used many times. You've likely got a particular gait -- one your friends recognize as you. These movements don't need to be consciously controlled; they're pre-programmed and automatic. Now suppose you hit some pebbles and start to trip. At this point, your brain intercedes and quickly adjusts and corrects your motor movement. Then a car suddenly careens towards you and you remember the words of your mother telling you to always walk against oncoming traffic; or it happens so quickly, you don't remember a thing and reflexively jump away from the road. Your heart is racing, your blood pressure's soaring and you've reacted even before you're aware of what's happened. But things regain a sense of normalcy and you're strolling along, taking in the children getting ready for school, the birds overhead, and the neighbors leaving for work. In an instant, you realize you're lost, and worse, you remember why you were taking a walk -- you were en route to pick up some bread and milk at the store. Now your brain goes to work: Should you retrace your steps? Do you have a general idea of where you are? Do you have your cell and who could pick you up? You start to feel anxious because you have lots to do today, so you decide to head to where the cars are rushing by, hoping that will lead you to a main street. And you're right. You get on the main road and begin to walk back towards your destination.

Now here is the neuroscience translation:

- While you're walking, one type of neuron -- your **sensory neurons** -- are taking in input from your senses which is sent up the **dorsal** (back, towards the skin) side of your spinal cord, up towards the

brain. **Motor neurons**, another type of neuron, send motor messages via the **ventral** side (front, towards your inner body) of the spinal cord, towards your muscles.
- Meanwhile, your **autonomic nervous system** is monitoring your heart rate, blood pressure, etc.; it's managing your internal **homeostasis**.
- Your preprogrammed movements have been learned by the brain. Pre-packaged, patterned movements are stored in the **spinal cord** which allows you to walk along, lost in thought about something unrelated to walking.
- The **sensory neurons** continue to take in information. Perhaps it starts to rain, or you feel cold and need to zip your coat. Your senses take in **internal** (inside your body) and **external** (from the environment) data of value to you.
- Your sensory neurons also provide **feedback** to the brain, in this case to the **premotor cortex** (by way of the **thalamus** and **somatosensory cortex**) of your brain. The **cortex** is also known in mammals as the **neocortex** or **cerebral cortex**. The premotor cortex is towards the front of the neocortex, an area known as the **frontal lobes**, and works alongside the **primary motor cortex** to make movement happen. Also, the **supplementary motor cortex** stores learned motor sequences (as does the cerebellum).
- The cortical areas of the brain work in conjunction with the **subcortical areas**, adding more subtle and nuanced thought and action to the human repertoire. This means that locomotion is generally handled by your spinal cord. The cortical parts of the brain provide "**conscious control**" or oversight.
- When you stumble, another part of the brain might chime in -- the **cerebellum**. The cerebellum learns and manages complex motor sequences. It plays an important role in walking and posture, and does so by integrating incoming information and then modifying "motor outflow" such that movement is smooth and organized.[1]

- The car careening towards you introduces the limbic system. When you recall the words of your mother, you are using your memory. The **hippocampus** is a member of the limbic system and plays a major role in the *creation* of memories. In this case, you're *recalling* something from childhood, from **episodic memory**. But if you don't consciously remember a thing, the **amygdala**, another part of the limbic system, is in charge. The amygdala stores emotionally significant, **unconscious memories**, especially those related to fear. As a result, you re-act: you jump out of the way of the oncoming car.
- When it suddenly dawns on you that you're lost, the brain must decide what to do:
  (i) The **anterior cingulate cortex**, also part of the limbic system, provides the recognition that you're off-course and need to take corrective action. The anterior cingulate cortex is in the front or anterior part of the **cingulate cortex**. The cingulate cortex is also called the **mesocortex** and acts as the controlling/coordinating part of the brain in creatures without a neocortex (and in some with…). Animals with a mesocortex are highly capable of managing their environment. Birds, for example, move beautifully, can locate and store food, and know to put off foraging when a predator is near. What's lacking is the more complex, subtle and especially *social* **behavior** the neocortex affords.
  (ii) In the anterior most part of the **frontal lobes** (which also include the premotor, primary motor and supplementary motor cortices) is the **prefrontal cortex**. Keep in mind, there are two of all of these regions -- one in each hemisphere of the brain -- despite the fact that they are often referred to in the singular. The prefrontal cortex gathers information from other cortical regions and uses **long-term memory** to decide what to do next. In the case of your walk, it sorts out where you are and your options for proceeding.

(iii) Meanwhile your limbic system, especially the amygdala, interacts with the **orbitofrontal cortex** (the mid-lower part of the prefrontal cortex) to give you that anxious, hurried feeling. The limbic system is a warning system; it uses emotions to guide behavior.
- Finally, your lateral prefrontal cortex maintains your **working memory** to reset and chart a new course. With the help of other subcortical areas, especially the **basal ganglia**, all other options are inhibited as you head towards the main road.

You now have a basic understanding of how the brain works: sensory input travels up the spinal cord; the brain turns it information and uses it to formulate a plan; and motor neurons send messages down the spinal cord to execute the plan of action. You also know actions tend to be "need-driven," and those needs can be either internal or external. What else?

1. The biology of the brain tells us that a general purpose of the brain is to allow the organism to manage its internal and external environment. One of the ways it does this is through movement: moving towards what it wants or away from what it wants to avoid. Also, that the brain works as a whole. While the neocortex adds an important layer of subtly and nuance, no region is in charge. No region is independent of the others. In fact, no single region does a complete function alone. While it's important to become familiar with the regions, what gives a region its function is "where the inputs come from and the outputs go to."[2] The brain is interdependent, synchronistic, interconnected, and mysterious.
2. The psychology of the brain tells us we only take in a piece of what's going on. In any given moment, there's a lot competing for our attention, so we only absorb what's important to us; we attend to some things and exclude others. In that way, we create our experience of reality: we can be walking aimlessly along one minute and in an instant, be rushed and anxious. We'll learn later

that the brain actually *constructs our reality*, as well as our experience of it. Similarly, a lot happens beneath our conscious radar. When we learn something new, the cortical regions of the brain are fully engaged in the learning process, but once it's learned, it's filed away and becomes unconscious. Conscious intervention is only required in novel situations or when something goes wrong.
3. The social nature of the brain teaches us that the motor system, and in particular motor skills *learning*, is a model for how **cognition** happens in the brain. Cognition is just another word for "knowing" in the most general sense. Some messages are "top-down," as from the prefrontal cortex to the muscles, while others are "bottom-up," as from sensory receptors to the brain. All happen in a context, in relationship with others.

And this is where the **wisdom** of the brain comes in. Just as coming face-to-face with a vicious dog can scare the life out of you, a heated discussion with a fellow employee or among clients can trigger fear, anger, and frustration. The brain reads the situation as a matter of survival – *social* survival. To the brain, **survival is first and foremost**. The brain manages the environment to ensure the organism survives and avoids death. As helping professionals, every part of the human ecology we attend to relates back to this core interest: it's why people are motivated by needs and desires; why they maintain a sense of self, possess a feeling of agency, and build a personal story; why clients imagine a future and recall the past, automate responses, and hold fast to their ideals; why we all need to feel in charge of our life, assemble a belief system, search for meaning and purpose, and want to belong. Leanings such as these are grouped into five over-arching human tendencies that appear throughout this book. They include:

- Our inherently conservative nature;
- The dominant role fear plays in our lives;

- How we build and live life around a self;
- The extent of human reactivity; and,
- The importance of knowing.

Each of these shapes our thoughts, feelings, and movements. And each harkens back to the reason we have a brain in the first place: to ensure the organism survives and focuses on living.

Recommended Reading
Neuroscience for Dummies, by Dr. Frank Amthor

A
Closing Thought

A Common Myth about the Brain: We only use 10% of our brain.

For this to be true, we would need to be able to remove a full 90% of our brain without any appreciable impact, which is simply not the case.[3] While some people have extra-ordinary abilities in some areas of their lives, they do not use more of their brain than other people. Also, while the brain can compensate for some level of damage or injury, a "repaired" brain is not wholly equivalent to an uninjured one, nor does an uninjured brain have neurons to spare; let alone 90% worth.[4] In a similar vein, there are no grandmother neurons. It was once thought that a single neuron could hold an entire memory, as of your grandmother. Since we don't lose entire memories as we age although we do lose neurons, this cannot be true.[5] Even in the case of dementia, what's lost is the ability to retrieve memories, not the memories themselves.

2

A Brief History

Neuroscience history is an interesting mix of anatomy, biology, and philosophy. Interest dates back to Egyptian times and the mummification process; to ancient Greece when the seat of intelligence was thought to be the heart; to Descartes' dualism which saw the mind and body as different "substances" linked together by the pineal gland; and extends up to today's network dynamics theories. One defining moment in early neuroscience involved two competing perspectives of the brain.[1] **Localism** saw the brain as organized around specific functions which could be physically located as regions mainly on the surface of the brain or **cerebral cortex**. Franz Joseph Gall, a physician and neuro-anatomist in the early 1800's, identified 27 such regions, responsible for faculties ranging from language to secretiveness, cautiousness, and hope. This contrasted with **holism** which conceived of the brain as an "aggregate field." To holists, the brain was a single circuit in which all brain regions, especially in the forebrain, played a part in every mental activity. As such, holists had little interest in studying the internal workings of the brain.

Gall's study led to the "discipline" of **phrenology** which sought to identify a person's personality traits and character based on the shape of his/her skull; it quickly lost credibility. During the same period however, a neurologist named John Hughlings Jackson proposed that a topographical map of the surface of the body -- called a **homunculus** -- was represented on each cortical area. This meant that the **somatosensory cortex**[2], for example, had a specific area designated for the forehead, adjacent to an area corresponding to the eyelids, the nose, lips, chin, neck, and so on. The existence of these maps was later validated, although even today there remains disagreement as to their precise organization.[3] Nevertheless, at the time, Jackson's findings gave a huge boost to supporters of localism. The localism perspective was further bolstered by the discovery of two brain regions specifically dedicated to language. In 1861 while studying brain lesions in stroke victims, Paul Broca identified what became known as **Broca's Area** in the left frontal lobe, the region responsible for speech. Whereas Broca's patient could understand speech but couldn't speak, in 1876, Karl Wernicki's patient could speak but didn't make sense. The resulting **Wernicki's Area** was responsible for language comprehension and meaning. Another neuro-anatomist, Korbinian Brodmann later organized the cortex into 52 distinct areas, many of which are still referenced today.

On a related front, researchers Ramon y Cajal and Camillio Golgi were debating the composition and internal communication of the brain. According to Golgi, the brain was a continuous mass of tissue. Cajal, however, used Golgi's staining method to prove that neurons are discrete metabolic units that communicate electrical information. Through the work of many scientists, the resulting **Neuron Doctrine** holds that: (i) neurons are the elementary building blocks of the nervous system; (ii) they process information via electrical signals; and (iii) those signals travel from **dendrites** to the **axon**. It was later learned that electrical signals travel *within* neurons and chemical transmission using **neurotransmitters** occurs *between* neurons.

Scientists now appreciate that the brain embodies both localism and holism: some functions are localized, while higher-order operations depend on connections that link multiple regions. And yet, even today, much neuroscience and related research centers on whether the mind comes from the brain working together as a whole, or from specialized components working somewhat independently.[4]

THE CONVENTIONAL MODEL OF THE BRAIN

In the 1970's, Dr. Paul MacLean introduced what may be the most frequently referenced model of the human brain: the **Triune Brain**. He divided the brain into three "evolutionary layers":

- The oldest, **reptilian brain** manages basic instinctive drives;
- The later, **paleo-mammalian brain or limbic system** envelops the reptilian brain and is key to memory and emotion; and,
- The most recent **neomammalian brain**, also called the **cerebral cortex**, supports awareness, thinking, and problem solving.

Iguanas, for example, have a reptilian brain. Reptiles are cold-blooded and mostly stimulus-driven, and while they can feel pleasure and pain, they do not generally care for their young, nor are they social creatures. Dogs have a reptilian brain that governs their instinctive drives, as well as a limbic system which is neatly covered over by a neocortex. As a result, dogs *feel*. Compared to a human brain, a dog's brain has less **prefrontal cortex**, meaning less capacity to pull together different kinds of information to support abstract reasoning and planning. While the evolution of the human brain has placed greater emphasis on vision, dogs have an acute sense of smell. Interestingly, the olfactory bulbs or so-called "primitive" organs of smell may be the evolutionary source of the more "advanced" structures of the human brain -- perhaps even the prefrontal cortex.[5]

MacLean's triune brain paints a picture of a brain built on top of a brain built on top of a brain. It demonstrates that evolution keeps what works: many of the older subcortical structures have been preserved rather than replaced; evolution is said to be more of a "tinkerer" than a designer.[6] The model also suggests that the "products" of evolution are generally well-suited to their environment. Natural selection makes choices for the good of the individual, the goal being to pass on the organism's genes. A given trait can be either good or bad depending on how well it helps the individual organism adapt to its environment.

While a useful starting point, MacLean's model has some serious limitations. First, evolution is not linear: it's not one smooth, logical progression with humans at the top. Evolution is not "an ascending scale. It is more like a branching tree"[7]; innovation can happen on any branch and there is no "top." Second, evolution is not additive:

one brain is not "lumped" on top of another, each performing its designated task. Rather, all three brains have continued to evolve with the addition of the next. The net result is a human brain that's an amalgamation of old and new structures; some older ones have been adapted to new purposes (called **exaptation**) while others have taken on altogether new roles.

There are other problems with the Maclean model. For example, the neuroscience community cannot agree on what comprises the limbic system such that the term has fallen out of fashion in many circles. But the most significant limitation of the model involves how the brain works: the brain is integrative, it works as a whole. New behaviors and cognitive skills evolved sharing a "basic set" of "neural structures," making the "division between primitive and advanced structures...totally arbitrary."[8]

THE HISTORY OF THE LARGER HUMAN BRAIN

Until recently, it was believed that the unique skills and abilities humans possess were attributable to the larger human brain. Evolutionary Anthropology and Evolutionary Psychology study the natural history of human beings and the human brain. Beginning with the generally accepted principle that all modern humans belong to a single species with a single origin, that being Africa, these fields outline the growth of the human brain over time, as shown below:[9]

Figure I
How the Human Brain May Have Evolved

Species	Approximate Time Line	Claim to Fame	Estimated Brain Size
Australopithecus	c. 3 million years ago	Walked upright, also comfortable in trees	About 450 cubic centimeters (cc), about the size of a chimp brain

Homo habilis	Between c. 2.5 and 1.5 million years ago	Used tools to eat their scavenger diet of meat, fruits, and nuts	Around 750 cc compared to our present size of 1,355 cc
Homo egaster Homo erectus	Between c. 2 million and 30,000 years ago	More human-like features; fashioned the classic "handaxe"	Around 1000 cc
Homo heidelbergensis	In Africa around c. 600,000 years ago, migrated to Europe over the next 200,000 years, died out c. 200,000 years ago	Hunter-gatherer mode of subsistence and potential for specialized labor, more cooperative hunting and living ventures [10]	Brain size close to ours, often attributed to meat-based diet
Homo neanderthalensis	c. 250,000 to 30,000 years ago	Distinct but coterminous species; best at surviving the 'ice age'	Brain size *larger* than ours -- perhaps by 15%! Cortical mass focused on vision v. frontal lobe

This brings us to anatomically modern humans, Homo sapiens, who are believed to have originated c. 200,000-100,000 years ago and moved out of Africa, replacing existing populations of Homo erectus, ergaster, heidelbergensis, and neanderthalensis. Such humans remained in close-knit groups of kin and their "own kind," as the interspersing of different races, etc. is considered a more recent phenomenon (c. 10,000 years ago).

As displayed in Figure I, the human brain increased nearly threefold over three million years. And yet, human technology did not keep pace with the growing brain: the first stone tools remained generally intact until the introduction of the hand-axe which again, changed little over the next million years or more.[11] Instead of a gradual progression, technological advancements occurred in leaps. Between 75,000 to 100,000 years ago, a cultural revolution (known as the Upper Paleolithic transition or Great Leap Forward) took place: tool technology ramped up and diversified; ornamentation became commonplace; art appeared in various forms; complex language skills evolved; and groups evolved both economically and socially.[12] Another, equally significant transition took place around 10,000 ago with the advent of agriculture, the introduction of food production and processing techniques, and the emergence of village life.[13] Both transitions suggest forces outside the brain's size were at work; as culture began to play a role alongside natural selection in shaping human evolution.

Since the larger human brain couldn't fully account for human advances, neuroscientists started looking at its component parts. For example:

- Over the course of evolution some regions of the brain have grown larger and others have reduced in size. There has also been significant functional reorganization in the human brain with many areas brain taking on additional, and often uniquely social, tasks;
- Humans have more association cortex than other primates,[14] and in particular what's called *multi-modal* **association cortex**, which allows us to drawn on and integrate information of various types, from multiple sources; and,
- The prefrontal area of the neocortex is larger in human beings than other primates.[15] The prefrontal region with its abundant connections gives us our capacity for abstract reasoning, planning, inhibitory control, higher level decision-making, and goal-directed behavior. Interestingly, other mammals -- echidnas are mentioned in one text -- also have an added piece of prefrontal cortex.[16]

It is now believed that it's not the size of the brain but its specialized circuitry that gives us our unique capacities.[17] Complex cognitive functions are possible because of the intimate and synchronized linkages between cortical and subcortical structures, and the dispersed, specialized networks that make up the modern human brain. Moreover, bigger brains and increased social pressures likely worked in tandem to build a better human brain: increasingly complex social relationships caused the need for ever more complex cognitive skills *and vice versa*. In his book, <u>The Making of the Mind,</u> Ronald Kellogg suggests the addition of five parts in particular led to a better brain, capable of leaps in technology and social living: working memory; executive function and in particular, planning and self-regulation; symbolic thought and language; the capacity for silent thought; and the ability to recall the past and imagine the future.[18] He argues that it's the interaction of these five parts that sets humans apart.

MODERN-DAY NEUROSCIENCE

Today's neuroscientists use fMRIs and related technology to study the brain, examining questions ranging from the effects of stress on the brain, to what the brain looks like during an insight and the neural bases of morality. **Functional Magnetic Resonance Imaging** (fMRI) is used to identify the "**neural correlates**" or "**neural signature**" of a given experience – to link the brain (the neural part) with the mind and body (the behavior, thought, or feeling being studied). fMRIs produce the color-coded, multi-dimensional images that show up in news stories and conference handouts, often with fantastic claims: "This is what the brain looks like in love!"

Magnetic Resonance Imaging (MRIs) and fMRIs take place in a scanner: MRIs examine the *structure* of the brain (the brain's anatomy), fMRIs look at brain *functions*. fMRIs are not a direct measure of neural events -- they measure the use of oxygen by neurons. The nervous system consumes around 20% of all the oxygen we breathe, and when performing a task, local blood flow increases to the areas of the brain that are active and the oxygen in that blood is consumed. The fMRI picks up the ratio of oxygenated blood to deoxygenated blood, called the BOLD (the blood oxygenation level-dependent response). To produce the color-coded pictures we often see requires several additional steps: researchers must create a baseline image, subtract the baseline from the "on task" image, use color gradations to plot confidence levels that the net area was not due to chance, use a computer to filter out any background noise, and then, finally, average the results of all the participants in the study.[19]

The biggest drawback of the fMRI is that it feeds the impression that an area of the brain specializes in a single function. An fMRI is often interpreted to mean that the area that "lights up" is responsible for that, and often only that, function. **Functional specialization** *is* a key organizing principle in the brain, especially in the cerebral cortex.[20] However, as we said, higher-order skills depend on complex networks that link multiple brain regions across cortical and subcortical areas. And specific brain areas are known to perform multiple functions. As a result, it's rare that researchers can crosswalk between a given mental activity and a specific area of the brain, and rarer still that a causal relationship exists: because your amygdala "lights up" when you experience fear does not mean your amygdala causes fear. The same area can also "light up" when you're happy or angry. There is at best, a correlation.

When it comes to using neuroscience research in studies *outside* the field, it still only makes sense to include it if the neuroscience adds something special. For example, if a researcher is studying the effects of meditation on decision-making, he/she should question whether including pictures of the "neural correlates" of meditation adds anything to the study or presentation. This contrasts with a study of what the brain is doing *while* meditating, in which case the neural correlates or "signature" would be essential. As a rule of thumb, unless the neuroscience offers a prediction, or identifies a biomarker, or in some other way provides a unique insight, it might be more of a distraction than an asset.[21]

HISTORICAL KNOW-HOW

This brief history points to the value of doubt. For helping professionals, the **wisdom** of the brain invites us to **regularly question** generally accepted assumptions and traditional models. In those instances when our options seem limited to either-or, it could be both… or a third option is waiting in the wings. As long as we question what we know to be true, we have a ready-made response to a standard refrain in the helping professions: a client says, "I don't know." The response is: "Yet…You don't know yet." Give it time.

A
Closing Thought

A Common Myth about the Brain: There are Right- and Left-Brained People.

The two **hemispheres** of the brain *are* different, but there are no left- and right-brained people. Rather than symmetrical, the human brain is "lateralized," meaning some functions are assigned more to one hemisphere than the other. The extent and nature of lateralization however, is still very much at issue. In any given presentation or conference, the two hemispheres might be displayed as follows[22]:

A Typical Portrayal of Left and Right Brain Differences

Left Hemisphere	**Right Hemisphere**
Logical Brain	Holistic Brain
Rational Analysis	Creativity, Art, Intuition
Generally Conscious	Generally Unconscious
Language	Emotional Aspect of Language
Thinks in Words	Thinks in Pictures
Coping	Relationships
Planning	Nonverbal Skills, Gut Feelings
Detail Oriented	Orients Attention
Positive Emotions	Negative Emotions

It is true, over the course of human evolution, the right and the left hemispheres have grown increasingly dissimilar. We also know that Broca's and Wernicki's area usually both reside in the left hemisphere, and handle speech production and language comprehension, respectively. This is called **hemispheric specialization**.[23] But language and speech are often used as the jumping off point for sweeping characterizations of two functionally distinct hemispheres: the left is logical and linear, processes information in a sequential or serial fashion and is less attuned to the physical body; the right is more visual/spatial, more emotional, intuitive, and creative, and processing occurs in parallel.

The right and left hemispheres are linked together by the **corpus callosum**, and no one function resides exclusively in one hemisphere. Even language is processed in both: speech is processed in the left and its emotional content is processed in the right.[24] Similarly, the encoding of certain memories relies more on the left brain, while retrieval relies more on the right. The same is true of creativity. In fact, *most activities* require both hemispheres. As a consequence, there are no left- or right-brained people -- there are only both-brained people, and the differences between the two hemispheres are more stylistic than substantive. Moreover, despite numerous attempts, there is no clean way to divide the two hemispheres that works in every case. We should therefore be wary of any theory or method predicated on the differences between two hemispheres or that emphasizes one over the other.

3

Hardwired Human Nature

The capacity of the human brain is striking: it controls *all* bodily functions while at the same time, allows us to interact socially, with other people as individuals.[1] The question remains, however, to what extent are humans *hardwired*? Today it's fashionable to refer to nearly every trait as wired into the brain: we're hardwired for willpower, for conflict, for love; to detect the intentions of others, for deception and cooperation; for belief, for God. But think about the brain's most essential task: survival is "wired" into the brain. Beyond that, it's a relative unknown. This has huge implications for the helping professions because the flipside of the hardwired question is: To what extent can people *change*? There are many ways to look at this issue. We will consider several of them.

HARDWIRED PATHWAYS AND INSTINCTS

In their book, <u>Neuroscience for Clinicians</u>, authors Simpkins and Simpkins describe four major pathways that are "wired" into the human nervous system[2]:

- *The Pain Pathway* manages the emotional and sensory aspects of pain and its moderation through natural opioid-like substances such as endorphins;
- *The Fear and Stress Pathway* mounts a brain-body response to a threat. This pathway produces "fight or flight," and, when chronic or sustained, fear turns to stress in this pathway;

- *The Reward Pathway* has two routes, both involve dopamine. One (via the **ventral tegmental area**) creates the sensation of pleasure and reward in response to sensory stimulation, such as having a massage. The other (via the **substantia nigra**) responds to motor movement which is why exercise feels good; and,
- *Other Regulatory Pathways* involve the regulation of sleep and the sleep-wake cycle, and the control of appetite.

A different approach to the hardwired question involves human instincts. Instincts are inherited. They are passed down through evolution and preprogrammed into our DNA because they work and increase survival.[3] In contrast, automatized behaviors are learned: we learn to ride a bike, dance the tango, etc. Instincts are "burned down so deeply into the circuitry" of the brain, we can "no longer access them."[4] For this reason, the idea of *human* instincts has fallen out of fashion because no one can agree on what's instinctive and what's not. Nevertheless, humans share certain "innate" qualities. Robert Winston, in his book <u>Human Instinct</u>, suggests the following list of instinctive or "innate" human qualities[5]:

1. We have a natural fear of snakes and crawly things, of heights, the dark, of blood and injury, and of dangerous situations. All of these date back to our ancient ancestors who survived longer because of those fears;
2. We seem to be "wired" for emotions, especially fear. In fact, we have an **early learning bias** towards fear;
3. We have certain inborn physical reflexes, and are "preprogrammed" for fight or flight, recognizing social dominance, and sex, eating, and sleep;
4. We are naturally inclined to make tools, manipulate objects, and mold/shape our environment; and,
5. We are innately social, can read facial expressions, have an inborn capacity for language, and need to play and explore.

Fight or flight has received a lot of attention lately. Every helping professional should know that fight or flight is a *biological* response. It can be due to a real threat or a perceived one. **Perception** is part of what the brain does; it's a complex process involving mental representations, stored memories, and numerous unconscious filters. It cannot be reduced to an "emotional brain" that takes over while a "thinking brain" goes offline. As we've said, the brain doesn't work that way; it works together as a whole. Moreover, the response is more properly known as **fight, flight, or freeze**: we fight when we're angry, flee when we're scared, and freeze when the danger is imminent and lethal. We will examine fight/flight/freeze in greater detail in later chapters.

The term "innately social" is also frequently cited in the media. Human babies are born helpless: to survive, they need to be fed and cared for by others. They have an exceptionally long period of dependency and learn *through relationships.* It takes other people to learn language, self-control, and cultural norms and in all those ways, humans are innately and necessarily social. But being "innately social" can also be used to imply we are hardwired for empathy, collaboration, and cooperation. We'll expand on this later.

HARDWIRED DRIVES AND HUMAN UNIVERSALS

Authors Paul Lawrence and Nitin Nohria identify four human drives as "hardwired modules" in the brain: the drive to acquire, to defend, to bond, and to learn.[6] The drive to defend is of special interest because it motivates so much human activity. The authors state that a perceived threat can be to a person's body, his/her possessions, bonded relationships, or "*cognitive representations*"[7] (emphasis added). In turn, our responses range from resistance to denial, rationalization, and counterattack.[8] Another "drive" that might be considered hardwired is our capacity for language. The issue surrounding this "drive" is whether it is a pre-formed, brain-based "module" or a function of "prepared learning." **Prepared learning** means certain behaviors are easier to learn during certain critical periods of development.

Other categories of "inborn" human qualities highlight our ultra-social nature. For example, we:

- Automatically assess the strength and trustworthiness of others;
- Attach to and attune with others, identify with a group and see our group as virtuous;
- Defer to and have an innate sense of social rank; and,
- Enjoy punishing wrongdoers and have an inner sense of fairness.

"Human universals" are traits shared by all peoples. Evolutionary psychologists have concluded that across cultures, humans need to be socially accepted and belong to groups; they want to influence others and protect themselves from harm; and they have a strong desire to form intimate relationships.[9] Other human universals include: prestige and status; property and inheritance; unequal distribution of power and wealth; the use of tools and myths, designated sex roles, and social groups; aggression and emotions, including hostility among groups and conflict within them; the use of gestures, grammar, and phonemes; kinship classifications, a taboo against incest, and particular facial expressions; and religion and the belief in the supernatural. [10]

A TOUGHER NUT TO CRACK

Obviously the hardwired issue is more complex than first thought. In his book, <u>Affective Neuroscience,</u> Dr. Jaak Panksepp outlines the primary "emotional operating systems" that motivate human behavior: the Seeking Systems, the Rage System, the Fear System, the Panic System, the Sorrow or Distress System, and the Lust, Care and Play Systems.[11] The first four are the major ones:

1. <u>The Seeking Systems</u> pertain to the pursuit of physical as well as cognitive needs and desires. They are aroused by the anticipation or expectation of a need or desire being met, rather than by consuming or the consummation of a desire. These systems motivate

us to take in the energy we need to survive and reproduce, and seek sources of energy that give life meaning. They are genetically prewired because of their survival value.

2. <u>The Rage System</u> motivates us to successfully compete for scarce resources. Anger is a by-product of the Rage System. Dr. Panksepp states that the most destructive forms of human aggression do *not* stem from the Rage System, but from our "higher brain" areas. He also notes that humans are "evolutionarily prepared" to "externalize the causes of anger and blame others," and anything that restricts our freedom will trigger anger, or worse.[12]

3. <u>The Fear System</u> is intimately linked to the Rage System. While the Rage System generates the "fight" part of the innate fight/flight/freeze response, the Fear System prompts the "flight" response. The Fear System generates fear-based anxiety and is said to induce a "powerful internal state of dread" in all mammals.[13]

4. <u>The Panic System</u> may reach as far back as the reptiles, to early distress mechanisms such as those that signal pain and cold. It later gave rise to separation-induced distress signals, like a baby's cry. This system generates panic-based anxiety. It is responsible for social bonding as well as feelings of loneliness, grief, sorrow, and loss. It is intimately linked to parenting and the brain-based systems that promote the care and nurturing of offspring.[14]

Panksepp's systems provide us with a framework for considering the extent to which human nature is "hardwired." He warns us though, not to expect too much: "Mature humans can voluntarily inhibit the expression of their primitive impulses and, with a great deal of social learning, can express their anger with the cool detachment of barbed words."[15]

THE BRAIN'S THREE SPECIAL PROPERTIES

Evolution has endowed the brain with three special properties. The first is that it's **embodied** – the brain lives in and through a body. Living is a full body experience of which the brain is only a part: the experience is visceral, sensory, cognitive, motor, emotional, social, cultural, and spiritual. Absent a body we cannot feel or sense or move. The body is where we experience agency; it's the source of our seeking behaviors. It's where resistance is expressed, where we feel what it is that we want, and where ideas, solutions, and new beginnings germinate. Everything we experience is the result of having a body.

The second special property of the brain is conservation. Taking some liberty, we can say the brain is **conservative.** It's a product of evolution and evolution conserves what's essential to survival. Like all living organisms, the brain conserves energy. It repeats what works; seeks to make things automatic; relies on shortcuts to avoid thinking too much or too hard; and prefers predictability. Because of its conservative nature, the brain craves *certainty.* The brain fills in gaps with guesses when real data is missing; prefers to maintain what is known in terms of knowledge, beliefs, attitudes, etc.; and believes the more familiar something is, the more likely it is to be true.[16] The brain's conservative nature is why we continue to pursue strategies that have long since stopped working, and tend to drift back to the status quo. Author Daniel Kahneman calls this the "**law of least effort**": given multiple ways to achieve the same goal, people eventually choose the "least demanding course of action"[17]...the path of least effort.

The third property of the brain is **emergence**. Emergence means the whole of the brain is greater than the sum of its parts. Consider a thought, for example...how it emerges; how the subjective quality of consciousness flows from physical matter; how perception appears and continues to change and evolve. The brain *itself* emerges – it is *self-organizing*, it literally builds itself. Aside from some genetic ground rules for where neurons go and the like, the brain works with the environment to form itself. Once formed, it regulates what it evolved from! It regulates itself through the autonomic nervous system, through self-regulation, and through social means as well, such as signage and police. This process of forming and reforming neural circuitry is called **neuroplasticity**. Emergent qualities are not inherently positive or negative; nor do they evolve and change in a linear fashion. The property of emergence allows us to create and change; to evolve as people. The **mind**, too, emerges...from energy and the flow of information within and among people.

HARDWIRED OR NOT?

Against the backdrop of an embodied, conservative, and emergent brain, how much does the hardwired issue really matter? The real problem with the concept of "hardwired" is that it's often misconstrued to mean fixed or immutable. The **wisdom** of the brain tells us otherwise. Just because something is hardwired doesn't mean it can't be changed. It means it's **harder to change**...because it has the force of evolution behind it. Overcoming the hardwired parts of ourselves takes *extra* effort because it's the direct opposite of Kahneman's law of least effort...

As you proceed through the book, it's important to keep the three special properties of the brain in mind. As a reminder:

- **Embodied**, the brain lives in and through a body. This also impacts our social relationships when we think of the importance of touch, and resonating with other people.
- **Conservative,** we default to the easiest path and conserve energy. We thus approach social situations and relationships with both ease and caution.
- **Emergence,** the brain and its activities are greater than the sum of its parts. This means we have some capacity to self-organize and regulate ourselves, as well as to re-organize ourselves and our relationships.

A
Closing Thought

A Common Myth about the Brain: It's Either Nature or Nurture.

The question, to what extent are humans hardwired, pertains to *how* we are. The nature vs. nurture question looks at *who* we are and the extent to which our genetic make-up or the environment is responsible. Genes can be thought of as the winners of the natural selection contest; survival of the fittest applies to the individual most able to survive and reproduce because of his/her successful genes.

Genes perform both a **template and transcription** role. In their template role, genes direct the organization of the nervous system, guiding the development and migration of neurons to their ultimate destination. There are 20,000 so-called **coding genes**. There are, however, 100 *billion* neurons in the brain; each typically connects with 1,000 other neurons at 10,000 different points.[18] Genes thus provide a template -- *not* the final design. The final design depends on plasticity mediated by the environment. The genetic blueprint is a potential -- the environment can either trigger these potentials to manifest (to "express" themselves) or not.

Let's take an example. Each of us has some potential for aggression. The genes relevant to aggression might include those related to pain thresholds, frustration levels, emotional regulation, and the like.[19] Whether we react aggressively depends on many factors in our environment: our upbringing, the level of violence in our neighborhood, our personal coping skills, etc. This is *not* to say there is a gene for aggression, rather there are genetic tendencies which can be triggered by the environment. As Dr. Sapolsky states: "Genes no more give commands than do telephone books"…and to understand them, "we must place genes in the context of the environment."[20]

Section II

How the Brain Works

4

Neural Pathways of Communication

To understand the brain, we have to understand neurons. Perception, thought, emotions, movement, memory, learning, self-control, social cognition are all by-products of how neurons work. The entire nervous system is built on neurons and nothing happens without them; in effect, "it's neurons all the way down."[1]

WHAT'S ALL THE FUSS ABOUT NEURONS?

The **central nervous system** allows us to take in our environment, interpret and decide what to do with that information, and then take action. It includes the brain, retina, and the spinal cord, and the **peripheral nervous system** which connects the central nervous system to the rest of the body -- to the muscles, organs, joints, etc. The **autonomic nervous system** is part of the peripheral nervous system and manages internal matters such as heart rate, body temperature, etc. There are two types of cells in the nervous system: neurons and **glial cells**. Glial cells are at least twice as numerous as neurons, and provide structure and support for neurons. Glial cells are involved in **myelination**, a process that creates a "coating" or insulation around the axons of neurons to make electrical signals flow faster. Recent research suggests a growing role for glial cells: **astrocytes**, for example, are a type of glial cell which are thought to participate in **synaptic plasticity**,[2] a process that strengthens the connections between neurons and serves as the basis for learning.

Neurons are special cells because they process information and communicate it to other neurons: "Neurons [are] the basic signaling units"... they take in sensory information, decide what to do based on some "relatively simple rules," and then, through "changes in their activity levels," pass that information on to fellow neurons.[3] The nervous system includes **sensory neurons** which process input from the senses; **motor neurons** which execute plans of action; local circuit or "**interneurons**" that process information in local circuits; and "relay neurons" that send information over long distances. Most neurons in vertebrates are neither sensory or motor but intermediate, or **computation neurons** – they take in input, compare it to the contents of memory, and devise a plan of action.[4] Some of the factors that make neurons special include:

- Neurons communicate with other neurons through signals;
- They respond to stimuli from both inside the body (e.g., a headache), and outside the body (e.g., a loud noise);
- Neural signals flow in one direction, from the dendrite to the axon;
- Most neurons do not touch. They are separated by a junction called a synapse which forms a gap. For a signal to move from one neuron to the next, it has to "cross the gap";
- Neurons act independently and collectively. The signal *within* a neuron is electrical. The signal *between* neurons is chemical – neurons use chemicals, called neurotransmitters, to "cross the gap";
- Neurons transmit information through the pattern of connections and excitation; and,
- Neurons generally do not reproduce; they grow from neural stem cells. In adult mammals, post- development, **neurogenesis** -- the creation of new neurons -- is very limited. New *synapses*, however, are generated.

Neurons have a cell body or soma, a surrounding **membrane**, intracellular fluid, dendrites, and an axon. **Dendrites receive** information from other neurons at synapses. Since dendrites are *after* the synapse, they are said to be "**postsynaptic**." Dendrites are tree-like structures that branch off the cell body. Depending on the degree of "arborization" (the extent of the branching), dendrites allow a neuron to receive inputs from many thousands of synapses. Motor neurons, for example, need to integrate multiple actions and therefore have extensive dendritic branching. **Axons send** information. Because they come *before* the synapse, they are "**presynaptic**." A single axon extends away from the cell body and has multiple branches with terminals at the tip that "synapse" on fellow neurons. For the signal to continue, it must cross the synapse. An axon may have thousands of **axon terminals** which synapse with the dendritic branches of other neurons.

Neurons come in a variety of forms. They can be unipolar with a single branch extending away from the cell body which includes both axon terminals and dendrites; bipolar (common in sensory processes) where the signal comes in one end at the dendrites and exits at the other, down the axon; pseudounipolar (a type of somatosensory[5] neuron) where the dendrites and axon have fused together; and the more familiar multipolar type which has many dendrites and one axon which then branches with terminals. Spinal motor and cortical sensory neurons, for example, are multipolar neurons.

NEURONS FORM PATHWAYS

Neurons form pathways of communication with other neurons to convey information. Because the process for doing so is complex, we'll begin with a snapshot of one neuron communicating to another. Snapshots don't happen in the brain; it's a continuous flow of information. Nor does a neuron receive or send a signal from/to a single neuron. Rather, it's bombarded by signals from other neurons and sends signals to many multiples of neurons. That said, let's look at Neuron B:

Step One   Neuron B is at rest.
Step Two   The dendrites of Neuron B receive a signal. The signal is such that it excites Neuron B, initiating a

change -- a depolarization -- in its membrane. If the change is sufficient to meet a given threshold, a spike or **action potential** is produced. The action potential is an electrical signal that is propagated along the axon.

<u>Step Three</u> At the end of the axon are many axon terminals which synapse on fellow neurons. Neuron B, at its "far end," is now presynaptic -- it's going to send a signal.

<u>Step Four</u> In order to cross the synapses to Neurons C, F, and K, Neuron B must release a neurotransmitter, a chemical which diffuses *across* the synapse and binds with receptors in Neurons C, F, and K (which at this point are postsynaptic because they are receiving the signal).

<u>Step Five</u> And the process begins again.

What follows is a more detailed description of each of the steps for how neural pathways operate:

Step One

- A neuron is surrounded by a "selectively permeable" membrane. The **membrane** itself is not porous but has thousands of **channels** that ions flow in and out of. An **ion** is a charged atom in a solution; in this case, the fluid inside or outside the neuron. Key ions in neural communication include sodium (Na+), potassium (K+), and chloride (Cl-), among others. The channels in the membrane either can open for specific ions (called "gated channels"), or are always open but permeable to only certain ions; hence, a membrane that is *selectively* permeable.
- When a neuron is at rest, the neuron is **negatively charged**, meaning that the environment inside the neuron is more negative than

the environment outside the neuron.[6] This difference in charge or **voltage** across a neuron's membrane is about -70 millivolts. This is called the "**resting membrane potential**" because it's a potential source of energy for the neuron to generate an electrical signal.

Step Two

- When a signal is received by a resting neuron, it causes a series of changes in the permeability of the neuron's membrane, meaning some channels open causing an influx of ions. If the change is positive and sufficient to meet a given **threshold**, an **action potential** is generates which flows down the axon to its **terminals**.
- Action potentials[7] are all or nothing propositions: if the signal is sufficient to meet the threshold, an action potential is triggered; the neuron is said to **fire or spike**. Spikes in axons are also self-generating, meaning they can carry a message over long distances without a change in amplitude. An action potential is followed by a short rest or **refractory period** before another action potential can fire.
- Triggering an action potential is a highly complex process that involves voltage-gated ion channels, the depolarization or hyperpolarization of cells, etc.[8] To learn more you can refer to this chapter's recommended reading.

Step Three

- At some point, the action potential reaches the terminals of the axon and the synapses that separate it from its fellow neurons. The neuron is now ready to relay the message to the next set of neurons. This once "signal-receiving" neuron has now become a "signal-sending" neuron; most neurons are both presynaptic and postsynaptic, meaning they both receive and send signals.

Step Four

- The action potential in the presynaptic neuron causes the neuron to release a chemical neurotransmitter from **vesicles** (pouch-like structures) that fuse with its membrane. The **neurotransmitter** then crosses (diffuses across) the gap and binds to **neural receptors** in the membrane of the postsynaptic cell. Once the binding occurs, any leftover neurotransmitter is "cleaned up" through a process called rapid uptake, or is diffused away or degraded by enzymes.
- Neurotransmitter targets are very specific; as opposed to say, a **hormone** which is released into the bloodstream and travels to sites throughout the body. The effect of a neurotransmitter once it binds to a neural receptor depends on the properties of that **receptor**. Most neurotransmitters produce either an **excitatory** effect or an **inhibitory** one. Glutamate and acetylcholine, for example, are neurotransmitters which act on receptors that are excitatory. GABA and glycine act on receptors that are inhibitory.

Step Five

- By this stage of the process, the receiving neuron is being bombarded with many messages from other neurons through its numerous dendritic branches; some are excitatory, others are inhibitory. At some point, these messages are summed together and if a **threshold** is met, the neuron "decides" to generate an **action potential**. It's always the sum of both inhibitory and excitatory messages that determines if the threshold is met. For example, self-control is *not* the sum of just inhibitory messages. While the act of controlling one's self may require inhibiting certain behaviors, *at the level of neurons*, there have to be enough excitatory signals to meet the threshold for an action potential to fire, otherwise the message – whether it's to stop or to start something -- never gets transmitted.

Even "at rest," some neurons continue to fire. This is sometimes referred to as spontaneous activity, or activity that's "**stimulus-*in*dependent**." It's thought that particular regions of the brain form "a coherent system that operates during the resting state" known as the **default network** or default mode.[9] The default network is "an anatomically defined brain system"[10] that we *default to* whenever we're not actively doing something else. Interestingly, the thoughts that dominate the default network always center on *our self and others* in relation to our self. The default network thus is an anatomically-defined, functionally linked brain network -- a "*physiological* baseline" which appears to be related to a *psychological* one.[11] It's an "internally directed" mode of cognition, not driven by the external environment, which we continuously revert back to and use for "remembering, considering hypothetical social interactions, and thinking about one's own future."[12]

The whole purpose of action potentials, neurotransmitters, crossing the synapse, etc. is to convey information. The message that's conveyed is not contained in the neuron or in the chemical makeup of the neurotransmitter, but in **the pattern of connections and excitation** that results. These patterns form pathways across neurons which if used repeatedly, ultimately coalesce into brain nodes, modules, hubs, and regions…opinions, characteristic gaits, recurring thoughts, etc. – into the **mind patterns** we use every day. And it all starts with neurons.

OUR SUBJECTIVE INNER STATE

Understanding neural pathways of communication is not just an academic exercise. They are laid down in the body – in *your* body. Our innermost needs for food, water, and the like, are communicated over neural pathways, enabling us to meet those needs through the environment. Sensory input travels over the same pathways and becomes perception, allowing us to experience the external environment. Together they create and sustain our **subjective inner state** -- where the inner and the outer worlds meet.

The subjective inner state begins *internally*, with the **viscera**. The viscera refer to the inner body, internal organs, and functions regulated by the autonomic nervous system. The brain regulates the body through the autonomic nervous system and hormonal releases. The viscera send and receive information concerning the body's state of **homeostasis**, the narrow range of body chemistry (such as temperature, blood sugar levels, oxygen, etc.) within which the body can survive. The viscera detect inner imbalances and communicate those needs using neural pathways. Building onto the viscera:

- **Emotions** are "autonomic physiological responses" to our needs[13];
- **Feelings** reflect our conscious awareness of emotions, the account the brain gives to explain them[14]; and,
- The outer environment provides the wherewithal to meet our needs.

The **wisdom** of the brain thus ensures **no one knows what it feels like to be you**. No one knows the level of chatter you hear in your head or what the voices say. No one knows how your body feels moment-to-moment, or what your wants, needs, and desires feel like inside. The subjective inner state is a very private world to which only the owner is privy.

Using those same neural pathways, we take in the *external* world through our senses, although it's not a *direct* transmission. The senses influence one another and are influenced by our expectations. When the body-brain interacts with a stimulus, our emotions and feelings are determined by our interpretation or **appraisal**, rather than the stimulus itself. Whether real or imagined, previous experience stored in the brain unconsciously shapes our interpretation. Thus what we perceive reflects "the active comparison" of sensory inputs with internal predictions[15]; in effect, our appraisal makes it so. Emotions and feelings take that appraisal into the body and tell us whether an essentially neutral stimulus is good or bad, a challenge or an emergency, a frustration or a dilemma -- and we feel and act accordingly.

Our subjective inner state is part and parcel of our **sense of self**. We each live life around a self; it is the center of our experience. Absent a sense of self there is no subjective inner milieu because there's *no one doing* the experiencing. Our sense of self and subjective inner state are nearly interchangeable with two possible exceptions: first, our sense of self comes with a feeling of **agency** -- a belief in, and the experience of, being able to act; and second, our sense of self carries with it a personal history, a **story of "me,"** and a commentator to narrate it. Our personal history is encoded into the brain and exerts a lifelong influence without our knowledge. New experiences have to be woven into that history – to be *made to fit*, such that we maintain a continually updated, coherent sense of self. For each of us, our inner world feels personal, private, and unique…our moods and ideas are far more real to us than are those of other people.[16]

Recommended Reading
Cognitive Neuroscience – The Biology of the Mind by Gazzaniga, Ivry, and Mangun.

5

The Human Nervous System

Neural pathways don't just exist in the brain. The entire nervous system depends on neurons and neural pathways. Every human capacity – from moving around, to touching someone's face, to sharing a good laugh – is rooted in neural connections and communication.

To understand the human nervous systems, we first need a way to refer to the body. The body is typically divided into dorsal/ventral and rostral/caudal axes:

- **Dorsal** usually refers to the top (think of a dorsal fin). In the special case of the spinal cord, the dorsal side is the side closer to the skin;
- Opposite the dorsal side is the **ventral** or bottom. On an animal, the ventral side would be its tummy or underside. In the spinal cord, the ventral side is closer to the interior body; and,
- The **rostral/caudal** axis runs front to back, from the face to the back of the head or head-to-tail.

Other terms include **lateral** which refers to the side, and **medial** which means the middle. A lateral view of the brain provides a side view; some medial views show the ***subcortical*** or interior parts of the brain, those covered over by the cerebral cortex.

AN OVERVIEW

The following overview of the nervous system includes the spinal cord, the brainstem, the cerebellum, the diencephalon, the limbic system, and the cerebral cortex.[1] During this review, it's important to remember that one region performs multiple functions, and one function depends on multiple regions. Also, that as one moves up the body, the information processed and the functions performed on it become increasingly refined. This allows us to analyze and respond to complex situations with a full suite of behavioral solutions.

1. <u>Spinal Cord</u>. The spinal cord serves as a special "relay tube" that sends and receives information between the body and the brain. The spinal cord takes in **sensory** information from neural receptors in the peripheral nervous system through the skin, muscles, and joints, and relays that information to the brain. It also sends **motor** information from the brain through motor neurons to the rest of the body to support both **voluntary** (you decide) and **reflexive** (automatic) actions.

    The connections of the spinal cord are divided into input **dorsal and** output **ventral horns**; dorsal, in this case, means towards the skin and ventral mean towards the interior body. The two horns organize the flow of information such that sensory input flows into the dorsal horn and motor information flows out through the ventral horn. Similarly, the nerve fibers that connect the spinal cord to the rest of the body include a sensory component (the dorsal root, fibers run from the skin and body extremities to the spinal cord) and a motor component (the ventral root, fibers run from the spinal cord to the muscles).

2. <u>Brainstem</u>. Above the spinal cord is the brainstem. The brainstem is fully functional at birth and has three major parts: the **medulla, pons, and the midbrain**. The brainstem regulates

many life sustaining functions; generates the energy necessary for fight, flight, or freeze; plays a role in certain aspects of hearing and balance; and includes a large section of the **reticular formation** which participates in arousal, pain regulation, and respiration.

3. _Cerebellum_. The **cerebellum** is a three layered structure that helps control movement. It is highly interconnected, meaning it **"projects to"** (has connections with) many other parts of the brain. Many of those connections are **reciprocal**, including those with the cerebral cortex.[2] The output of the cerebellum is believed to be "entirely inhibitory": it adjusts or modulates behavior, controlling its "rate, rhythm, and force"...in effect, it's _quality_.[3] According to recent research, the cerebellum also participates in cognition, language, memory, and reasoning.[4]

4. _Diencephalon_. The diencephalon is comprised of the thalamus and the hypothalamus. The **thalamus** is the link between the cortex and the rest of the brain. With the exception of olfaction (smell), all signals from the senses and the subcortical regions _to_ the cortex, and much of the communication among the various areas _within_ the cortex, pass through the thalamus. More than just a relay station though, the thalamus controls _which_ sensory information reaches the cortex, thereby integrating incoming and outgoing information which ultimately affects memory, motor movement, and goal selection.[5]

The **hypothalamus** regulates our internal states. It's involved in homeostasis, the circadian rhythms of sleep and wakefulness, the regulation of thirst and appetite, and the release of hormones by the pituitary. The hypothalamus plays an important role in the autonomic nervous system and the endocrine system, and is part of the **hypothalamus-pituitary-adrenal axis** (HPA) which participates in the body's stress response. Through the endocrine system, key hormones related to the hypothalamus include

vasopressin and oxytocin which will be discussed when we consider how parenting shapes the brain.

5. <u>The Limbic System</u>. According to **Cognitive Neuroscience**, **cognition** is the process of:
   - Taking in sensory information as *perceptions*;
   - Using memory to assemble those perceptions into "*representations*";
   - Translating those representations into *ideas and feelings*; and,
   - Turning those ideas and feelings into *action*.

Cognition involves thought, memory, and emotion; they are intertwined. The **limbic system** works with the **orbitofrontal cortex** to integrate emotions and memories into thoughts and ideas.[6] It performs an "evaluative process," helping to discern what events mean and decide whether they warrant our attention.[7]

While the term is falling out of use, the limbic system remains a useful reference tool. Most structures of the limbic system are **subcortical** -- they are tucked underneath the cerebral cortex which folds around them. This applies to the orbitofrontal cortex as well as the **orbitomedial prefrontal cortex**, both of which are part of the **cerebral cortex** but are active players in the limbic system. Limbic structures are also extensively connected with cortical and other areas, making the creative, decisive, and abstract work of the cerebral cortex possible. Additionally, the functions of many of these regions have a distinctly social flavor.

A list of limbic structures usually includes the amygdala and the hippocampus; the cingulate cortex (or cingulate gyrus); and the basal ganglia which is comprised of the striatum, substantia nigra, and other areas. In evolutionary terms, the **amygdala** and the **hippocampus** add memory to behavior and its control.[8] As a result, a rabbit can remember what it means that a hawk in flying overhead, despite the fact that it's hungry. The **cingulate cortex** serves as the neocortex for many creatures. You may recall its frontal most area, the **anterior cingulate cortex**, provided the realization you were off course in chapter 1 and needed to take corrective action. The evolutionary origins of the cingulate cortex extend back to the appearance of maternal behavior; it is thought to be part of the neural "infrastructure" for social cooperation.[9] The **basal ganglia** perform a variety of functions, including helping to initiate and control movement. The **insula** should also be mentioned here because it has extensive projections (connections) with many limbic structures and, working with the cingulate cortex, allows us to know how we feel internally and on an external emotional level.[10]

6. <u>Cerebral Cortex</u>. The cerebral cortex has two **hemispheres** and four main lobes which are tied together by the **corpus callosum**. The cerebral cortex is surrounded by collagenous fibers and the skull for protection. Its massive numbers of neurons are arranged in folds, allowing its large surface area to be efficiently packed into the skull while cutting down the distance axons must travel to conduct their signals. The folds include **gyri** or protruding areas, and **sulci** or **fissures** which are "valleys." A single neuron in the cortex is likely to receive inputs from multiple neurons located throughout the brain. The surface layer of the cerebral cortex appears gray because it is densely populated with **cells bodies and dendrites** that receive information. Beneath the surface is a thick layer of white matter representing billions of **axons** that appear white because they are **myelinated** to make electrical signals flow faster. The corpus callosum is also predominately white matter. The axons serve as "cabling" to connect the cerebral cortex with other cortical and subcortical areas of the brain, and tie the two hemispheres together.

The **neocortex** comprises a full 90 percent of the cortex. The neocortex is a 6-layered structure that averages around 3mm thick and encases the subcortical regions of the brain. It gives us our capacity to plan and execute behavior, to perceive, act, think, emote, and remember. The neocortex has multiple columns and is populated with **pyramidal cells** which project over long distances, as well as other types of neurons. Each layer of the neocortex performs a distinct function[11]:

- Layer I is closest to the surface and composed mostly of dendrites and some axons;
- Layers II and III connect local clusters of neurons within the same cortical area – they "project locally" -- as well as to other cortical areas. These layers manage *intra*-cortical communication;
- Layer IV receives input from the thalamus;
- Layer V projects to other cortical and subcortical areas (especially the thalamus), and provides the "major output pathways" of the cortex; and,
- Layer VI is closest to and blends into the axonal cabling of the white matter which lies beneath it.

The neocortex is organized into four major lobes:

- The **occipital lobe**, at the back of the brain, manages visual processing;
- The **temporal lobe**, above the ears, combines visual and auditory information, also higher order vision such as object and facial recognition;
- The **parietal lobe**, near the top of the brain, includes the somatosensory, gustatory, and parietal association cortices, and helps manage attention and spatial representation; and,
- The **frontal lobe**.

The frontal lobes include the **primary motor cortex**, the **premotor and supplementary motor cortices**, and the **prefrontal**

cortex. The frontal and prefrontal cortices handle both behavioral and emotional matters; the information they receive comes already highly processed by other parts of the brain. The remainder of the neocortex, which is neither sensory nor motor in nature, is sometimes called the association cortex, although this term is not really accurate because most of the cortex has some specific function. Different association areas pull together information from different sources across the cortex and perform functions that are "multi-modal" (involving multiple senses) in nature.

The **prefrontal cortex** can be broken down into the **dorsolateral** (top-side) **prefrontal cortex** (DLPFC) and the **orbitofrontal** (front-center) **cortex** (OFC). The dorsolateral prefrontal cortex focuses more on working memory and thoughts, while the orbitofrontal cortex is more involved in emotions and providing emotional guidance. Together they produce **goal-directed behavior** -- the behavior designed to meet our needs. It's often said that the primary role of the prefrontal cortex is inhibition; that its main job is "executive control." But the prefrontal cortex is not the "rational brain," it does not exert control over our emotions and ensure we do the right thing. The prefrontal cortex allows us to act in a goal-driven manner, to be *intentional* – and intentions can be good or bad.

PATHWAYS THAT CHANGE

The neural pathways of communication in the human nervous system can be thought of as an idea, a thought, or a series of movements. Consider the act of hitting a tennis ball: the **prefrontal cortex** has the idea of hitting the ball; it projects to the **motor cortex** of the **frontal lobe** sends a signal to your **spinal cord** to move the muscles in your arms and legs; the **basal ganglia** selects among the movements you've learned, while the **cerebellum** adjusts and refines your movements and balance; the posterior **parietal cortex** gives you a continuous sense of where your body is in space and where your racket is vis-à-vis your body; your **occipital cortex** keeps track of the ball and your fellow players; and your **brainstem** manages your heart rate and respiration.[12] But what happens once these pathways are laid down? How are they improved upon, updated, or changed?

Neuroplasticity is the way the brain alters its neural structure in response to experience. It's how we incorporate what we learn into the brain and into our lives. Neuroplasticity is why the more you practice your tennis swing, the more it improves: it improves because the connections in the pathways grow stronger. Neuroplasticity has been extensively studied in the cortex. **Neurogenesis** is the creation of new neurons which, according to current knowledge, is limited in adults to two regions of the brain, the hippocampus and the olfactory lobe. While neurogenesis continues throughout life, it is restricted to two kinds of neurons and to those two regions, as far as we know.[13]

Synaptic plasticity and **synaptogenesis** are the more common ways experience gets incorporated into the brain. As opposed to creating new neurons, these processes apply to *existing* neural pathways. Both increase the strength of connections between **synapses.** Synaptic plasticity increases the strength of connections through repeated use – as in the case of your tennis swing. Synaptic plasticity is faster and easier to invoke than synaptogenesis. Synaptogenesis involves the creation of *new synapses* in existing pathways. It does so mostly by adding new spines (little extensions) to existing synapses, increasing the number of potential "contact points" in a given pathway and thereby increasing its strength. Synaptogenesis may result from "sustained plasticity."[14] When synaptic plasticity and synaptogenesis mediate learning and memory in the hippocampus, this occurs through a process called **long-term potentiation** (LTP). Long-term potentiation involves repeated and simultaneous firings at multiple sites along the neural pathway. If this takes place within a given time frame, the memory or behavior is more likely to "stick." If not, it fades away and is lost. We'll talk more about long-term potentiation when we discuss learning and memory.

Neuroplasticity allows us to grow and change – it literally changes the neural pathways in the brain based on life's experiences. In this way, the brain can continually make sense of its environment. But neuroplasticity is not always positive. We can learn and retain horrible things -- thoughts and memories that narrow our perspective and shut us down, and it's not always under our control or a real choice. Through neuroplasticity, the **wisdom** of the brain ensures we have **the potential to change** – to repurpose and revise ourselves, and evolve as people…*within reason.*

6

The Organization of the Brain

We now know neurons communicate with other neurons and form pathways throughout the nervous system. Once pathways are laid down, similar pathways form nodes and clusters; communities of clusters become modules and hubs; like-minded modules create regions; multiple regions work together; and all are tied into large complex networks that crisscross the brain and link functional areas, giving rise to perception, cognition, and action. To accomplish this, the brain is built and organized around several key concepts:

- *The brain is rooted in survival, driven by needs, and characterized by competition.* At the heart of the nervous system is the survival of the organism; to keep it alive and focused on living. The very first movement is approach and withdrawal: to move towards food and safety, and away from danger. Movement towards and away is evolution's reason for neural communication; it may be the original duality. Competition is another key organizing principle of the brain: there is competition for space, resources, and survival. During development, there is competition among neurons for synaptic connections: in the sensory system, for example, visual, auditory, somatosensory, taste, and olfactory neurons compete for available synapses. Similarly, since only a small group of neurons can be active at one time, clusters of neurons "constantly compete for limited attentional resources."[1]

- *Functional specialization is an essential feature of the brain, especially in the cerebral cortex.* That means *all* of the following statements are true:
  a. Brain regions specialize in functional areas.
  b. Functional connections can exist among brain regions that are not anatomically linked.
  c. The same region can process both cognitive and emotional information.
  d. Complex skills depend on many distinct brain regions.
  e. It's the connections between regions that make complex skills possible.
- *The brain is self-organizing: it builds itself and then manages that from which it evolved.*
  (i) Through genes, natural selection provides a template for the organization of the nervous system. The final design depends on the environment which can trigger a given trait to be expressed by changing the structure of the nervous system.
  (ii) A trait can be an **epiphenomenon**, meaning it is the "by-product" of another, naturally selected trait. For example, some believe **consciousness** is an epiphenomenon: rather than deliberately "selected for" in evolution, it may be a by-product of our complex neural circuitry.[2]
  (iii) The brain includes many structures that have undergone **exaptation**, where a given function has been adapted to serve a different or expanded function.[3]
  (iv) Through plasticity and emergence, the brain grows and changes in response to experience. It *learns* in order to keep pace with its environment.
- *Functional areas of the cortex are close to one another.* Evolution has ensured a minimalist approach to wiring in the brain. Since it wouldn't be efficient for every neuron to be connected to every other neuron (called "universal connectivity"), each neuron is

connected to, on average, several thousand other neurons and *the densest connections are close by* (called "**small world architecture**").[4] This conserves space by limiting the amount of cabling required and reduces the time it takes to conduct neural signals.

- *Neurons that fire together consistently over time, ultimately wire together.* This is the basis of what's called **Hebbian learning**: "Neurons that fire together, wire together." The maximum number of connections or synapses among neurons (called **peak density**) occurs during the first 15 months of life.[5] This is followed by a period of **pruning** (synapse elimination) and then stabilization, which can last up to a decade. But **synaptic plasticity** continues and the more a given pathway is stimulated, the stronger it gets (See Chapter 5). Circuits that are not used die off. We each have many **mind patterns**, repeatedly reinforced patterns of thought or behavior that over time, become rules, assumptions, expectations, beliefs, judgments, opinions, actions, etc. We automatically return to our mind patterns; they are part of the brain's conservative nature.
- *Brain networks support functions that are shared among humans.*[6] There is enormous variability among individuals -- in genetic attributes, numbers of dendrites, amounts of and sensitivity to neurotransmitters; different thresholds of neural excitability, patterns of connectivity among neurons, brain structures and anatomy, and uniquely different network topographies. Despite this, "human brain networks support behavioral and cognitive functions that are, for the most part, shared among all individuals."[7] Referred to as **functional homeostasis**, it's been suggested that this occurs because network regulation happens at a global level rather than locally, which allows the brain as a whole to remain stable in the face of constant change.[8]

REGIONS OF THE BRAIN

When talking about brain regions, we run the risk of learning to associate a particular region with a specific function. Let's take an example. Assume you believe the hippocampus "does memory." In fact, the hippocampus plays a major role *in the creation* of memory.[9] The amygdala also "does memory," it works with emotionally-charged memories. In addition, there are many different kinds of memory, such as episodic, semantic, etc. Moreover, the hippocampus does not *store* memories; it encodes explicit memories and they are stored elsewhere in the brain.

Just as importantly, by looking at the hippocampus in isolation we forget where learning and memory come from: absent an environment, there's nothing to learn and no memories to store; absent a body, there is no means to experience the environment or create those memories; and absent a brain, there's no past to compare and contrast an event with and reflect upon its meaning. The **wisdom** of the brain is that it is immersed in a **brain-body-environment interaction**, the product of which is the experience we have of ourselves in the world around us. With this in mind, we will consider several regions of the brain in greater detail. In doing so, take note of the range of functions ascribed to each; the breadth of connections – where each projects to and what projects to it; and how each region links the brain and body to its environment, giving us a sense of purpose, relationship, and control.

Figure I
Limbic Structures

Region	Select Functions	Interesting Facts
Limbic System	Includes emotion, fear conditioning, fight/flight/freeze response, learning and memory	Adds memory to instinct

Basal Ganglia	Plans, organizes, and executes voluntary bodily movement and its coordination Works with prefrontal cortex to choose among goal-directed activities and inhibits interference once a choice is made	Functions are not well-known, is sensitive to reward qualities of the environment Sees patterns from one situation to another
Amygdala	Generates, processes, regulates, and helps control a broad range of emotions Helps manage environmental threats and challenges Stores memories with emotional significance for use in the same or similar situations **Amygdala/ Orbitofrontal Cortex**: Learns about appropriate behavior to avoid negative consequences; creates an inner bodily feeling of unease	Attuned to fear, which may be our most powerful early memory Involved in fear conditioning, producing unconscious, autonomic responses Provides complex emotional, cognitive, and sensory input to survival issues

Hippocampus	Especially involved in the creation of declarative (explicit) memory; in transferring short-term into long-term memory Encodes episodic memory linking one's self to one's environment Facilitates the integration of new information and previous learning **Hippocampus/Lateral Prefrontal Cortex**: Recalls external events without stimuli being present	Evolved prior to neocortex, possibly from early systems used to navigate the environment Can grow and shrink depending upon experience

Figure II
Insula and Cingulate Cortex[10]

Region	Select Functions	Interesting Facts
Insula	Connects inner bodily states with outer experience Involved in many basic emotions and may be associated with emotional and self-awareness	Triggered by facial expressions, feelings of mis-trust, changes in eye gaze, frustration, and pain

Cingulate Cortex	Involved in motivated, creative, and/or spontaneous behavior, and helps focus attention Plays a role in maternal attachment, the emotional reaction to pain, and control of aggression	Also known as Mesocortex, also the Cingulate Gyrus First appeared in animals exhibiting maternal behavior, which may be a source of social behavior
Anterior Cingulate Cortex	Monitors progress towards goals – detects errors, allocates attention, and monitors responses Monitors and integrates the activity of many other brain areas Involved in the experience of physical pain and social pain (dorsal area)	Has extensive projections to the LPFC Plays a role in deception

Figure III
Cerebral Cortex

Region	Select Functions	Interesting Facts
Cerebral Cortex	Coordinates perception, thinking, reasoning, language, speech, visual processing, and memory Depends on multiple lobes and extensive subcortical connections to do its job	Adds complexity and nuance to analysis, alternatives, strategies, and decision-making Complements a brain that can already take appropriate action within a given environment
Parietal Lobe	Receives sensory information about touch, temperature, and pain in or on body – creates the experience of a "somatic self" Situates the body in space and assists in processing visual-spatial information	Contains the somatosensory, gustatory, and parietal association cortices
Frontal Lobe	Involved in planning and organization of action, parts of language, control over movements, and social judgments	Evolved to organize complex motor movements Language may have co-evolved with fine motor movements

Temporal Lobe	Processes auditory and visual information from the environment and aspects of language Includes the **fusiform face area** (in medial part of lobe) that processes faces	Left hemisphere plays a larger role in spoken language
Occipital Lobe	Processes visual information from environmental input	Creates a 3-D picture from a 2-D image
Corpus Callosum	Links the two hemispheres of the brain	Plays a unifying role in the perception of objects
Prefrontal Cortex (PFC)	Coordinates abstract reasoning, executive functions, and directed attention	May have evolved from the olfactory bulb Includes the dorsolateral prefrontal cortex, orbitofrontal cortex, orbitomedial prefrontal cortex, and medial prefrontal cortex, among others
Dorsolateral Prefrontal Cortex (DLPFC)	Involved in executive functions such as complex goals, planning, attention, and problem-solving Key to working memory and essential for episodic memory	Processes independent thinking that is not environmentally-driven

Orbitofrontal Cortex/ Orbitomedial Prefrontal Cortex (OFC/OMPFC)	Adds emotional cognition and bodily state awareness to the situation at hand, is involved in risk and reward assessment and moral judgment Plays a role in empathy, self-control, and navigating social situations Heavily connected to amygdala	Receives input from all the senses Integrates social information to help guide perception, action, and interaction
Medial Prefrontal Cortex (MPFC)	Involved in mentalizing about others and conceptualizing about the self Along with **Septal Area**, is involved in empathy	Works with **temporoparietal junction** (TPJ), **precuneus**/posterior cingulate cortex, and **temporal poles**

THE PREFRONTAL CORTEX IN GREATER DETAIL

The prefrontal cortex (PFC) has been called the executive brain, the rational brain, even the thinking brain. Situated in the frontal most part of the frontal lobes, the prefrontal cortex is extensively connected: it receives **afferents** (inputs) either directly or through the thalamus from the brainstem, the pons, the hypothalamus and sub-thalamus, the amygdala, the hippocampus, other limbic regions, areas related to an organism's internal needs and motivational states, the cerebellum, and the ventral tegmental area (where dopamine and serotonin originate), and is abundantly connected with other cortical regions.[11] It is because of these abundant connections that the prefrontal cortex can perform complex and abstract functions such as planning, higher level decision-making, analysis and abstract thought, inhibitory control, and goal-directed behavior.

The prefrontal cortex serves as point of convergence and association.[12] Highly processed and refined sensory, motor, visceral, and other input flow into the prefrontal cortex where it is merged, associated, and integrated. The prefrontal cortex blends multiple inputs and multiple senses to shape ideas, goals, and behavior. It is among the last neural systems to evolve and the slowest to develop.[13] The prefrontal cortex:

- *Manages our attention.* Attention requires **top-down** pathways (as in descending, generally controlling or inhibiting) because it is voluntary; and **bottom-up** pathways (as in ascending, more direct or spontaneous) because attention is often stimulus-driven or related to a bodily need or desire. Since the brain can only pay attention to a sliver of what's going on, the prefrontal cortex helps direct and focus attention;
- *Participates in working memory.* **Working memory** keeps our attention trained on one or more thoughts. It is a "dynamic pattern of neural firing" that's temporarily maintained by the prefrontal cortex, integrating multiple ideas in a "recurrent loop."[14] So long as we keep replaying the loop, we maintain the thought and can do so without any stimulation from the outside; because it's all in our head. But if we fail to hit the replay button, we lose the thought. Humans have roughly the same working memory capacity as a crow, although crows think in terms of items and humans deal in chunks of items.[15] Working memory is supported by the dorsolateral prefrontal cortex and the hippocampus. It's key to how we organize and prioritize our thoughts, ideas, decisions, plans, and actions;

- *Integrates thoughts and emotions, and helps regulate emotions.* The prefrontal cortex adds "subtlety, complexity and adaptability, and perseverance" to our otherwise instinct-driven pursuits.[16] The dorsolateral prefrontal cortex provides intention and focus, while the orbitofrontal cortex provides emotional guidance. When it comes to managing emotions, the prefrontal cortex shares its authority with the amygdala. The prefrontal cortex has top-down projections from the orbitofrontal cortex to the amygdala and other limbic regions, such that it can overrule the amygdala -- but it takes effort.

Everyone loves the prefrontal cortex. Some helping professionals talk about "snagging the prefrontal cortex" in the hopes of bringing its deliberative, logical, and mindful qualities to teaching, mediation, therapy, and treatment. They advocate a top-down approach, where the prefrontal cortex asserts control over and inhibits the more "primitive" parts of ourselves. **Neuro-reductionism** is a view that argues higher order mental processes stem solely from the biological properties of neurons and their patterns of connection. **Neuro-centrism** holds that human behavior is best understood from the "predominant or even exclusive" perspective of the brain.[17] **Cortico-centrism** is the belief that the cortex, and especially the prefrontal cortex, is all that matters.[18] Leonard Koziol and Deborah Budding wrote an entire book refuting cortico-centrism. They suggest that not only is the prefrontal cortex abundantly connected, but many subcortical regions are actively involved in cognition and some actually guide and modulate the prefrontal cortex to act in a particular way.

The prefrontal cortex is a limited resource. We already know that it is more *intentional* than rational (see Chapter 5). By sharing its authority with the amygdala, the speedy amygdala can easily overrule the more discriminating orbitofrontal cortex, making it difficult to control ourselves. What's more, the prefrontal cortex uses a lot of energy and tires quickly from overuse. It's readily distracted as we often revert back to the **default network**, thinking about self and others. And it reflects the brain's conservative nature: the prefrontal cortex prefers to work in a controlled, calm, stress-free environment, doing one thing at a time. Like so much of the brain, its connections are **experience-dependent**[19]: as opposed to an executive, rational, or thinking brain, the capacity of the prefrontal cortex depends…on life experience.

7

How the Brain Might Work

Neuroscience is often described as the study of how the brain supports mental processes – *of how the brain works*. Most of the research is firmly rooted in neuro-reductionism[1] which views the brain and the mind as one; where the brain *gives rise* to the mind. This can lead to questions like, "Am I my neurons?" Or, "Do my neural connections fully account for who I am?"

There are many models of how the brain works and most share certain features in common. One is **modularity**. Modularity refers to the way clusters of neurons (as opposed to regions) specialize in a particular function – for example, how clusters in the vision system detect the edges of objects. Each module collects only its piece of the data set and fires in response to highly specific criteria. In addition, modules are often anatomically distinct. To work, modularity-based models require an overriding independent module to integrate the output of fellow modules and orchestrate higher order systems. This task is typically assigned to the **prefrontal cortex** which is given the role of "central executive," making the decisions necessary to transform information into thoughts, feelings, and actions.[2]

Another common feature is **hierarchical processing**. When the brain processes information, raw data flows from one tier to the next and on up the line, as the data becomes increasingly filtered and refined into useful information. Information is processed **serially** and in **parallel**: "serially" means each area of the brain performs a given task on the data and then passes it off to the next processing unit in the chain; "in parallel" means that the same information is "processed differently in parallel pathways."[3]

One other feature shared among models is the idea of **"parallel distributive networks."** By now you know a perennial problem in neuroscience has been the desire to assign a specific function to a particular part of the brain, or an entire region to a given function. This has been largely unsuccessful because that's not how the brain works: the more complex a behavior or operation is, the less straightforward the so-called **"structure-function relationship."** The thinking behind parallel distributive networks is that for the brain to support cognition and other complex functions, it must rely on systems that work in tandem (parallel) and draw on capacities across the brain (via distributive networks). In other words, the more complex the function, the more widely distributed the systems are that support it.

Against this backdrop, this chapter considers two different models for how the brain might work. The first model is the *computational theory of mind* which sees thinking as a computation. The second is the *network dynamics* approach. No one knows for certain which aspects of what model are correct or which will become pieces of future models. Regardless, the **wisdom** of the brain is that despite myriad pathways and multiple regions, the brain-body-environment interaction **produces a single, unified picture of reality** which we live in and through our entire lives...often without ever questioning it.

THE COMPUTATIONAL THEORY OF MIND

The computational theory of mind regards thinking as the sum of many series and levels of computations. According to this theory, complex behavior can be broken down into manageable parts: the brain takes in data inputs, performs computations on those inputs, links inputs with symbols or representations stored in the brain (a topic we will return to in Chapter 9), and translates the results into thoughts and actions. At each step:

- Information is processed hierarchically, progressing from lower to higher levels of processing within and among cortical areas;
- Processing occurs both serially and in parallel across distributed networks;
- Massive computations are performed at each step; and,
- Information becomes increasingly filtered, refined, complex, and integrated.

The net result is **cognition**.

In his book, How the Mind Works, Steven Pinker describes how the computational theory of mind envisions brain activity as akin to a set of cognitive programs: the programs are "assemblies of simple information-processing units -- tiny circuits that can add, match a pattern, turn on some other circuit, or do other elemental logical and mathematical operations."[4] The content of those programs depends on the neural pathways of communication. A "computation" in this case refers to the transformation of physical energy from a stimulus into patterns of activity among neurons; that activity then spreads throughout populations of neurons, through memory and association areas, and ultimately builds to a picture, thought, or plan. According to this model, it's the properties of the neurons and the networks of which they are a part that account for how the brain works. By the time the product reaches consciousness, the multiple streams of information and levels of computation meld into a seamless, coherent picture of the world around us.

Perception provides an example of the computational model.[5] Perception is the product of the senses; it's our experience of the outer and our inner world. Each of the five senses has a system in the nervous system dedicated to processing its information and they work in similar ways. Sensory information is taken in through **sensory receptors** in the body: **mechanoreceptors** detect when skin is stretched or pulled, motion in the inner ear, extension in the bladder, etc.; **chemoreceptors** receive information about taste and smell; **photoreceptors** are involved in vision; and **thermoreceptors** deal with pain and other visceral matters. Each receptor responds to a single class of stimuli, at a specific location, and fires in a unique way depending on the type of stimulus energy it receives. In so doing, the receptors *transduce* (change) the energy of given stimulus – sound waves from the environment, for example -- into a workable signal that flows along particular pathways in the brain and ultimately becomes perception.

The spinal cord takes in *sensory* information through sensory receptors in the skin, muscles, and joints. The input flows up to the **thalamus** which, with the exception of the olfactory system, links sensory receptors to the cortex (See Chapter 5). The spinal cord then sends *motor* information from the brain via motor neurons to the rest of the body to perform voluntary and reflexive actions. In the **cortex**, each of the major processing areas includes one or more **maps** displaying the parts of the body. Called a **homunculus**, the **sensory homunculus** includes four maps of the surface of the skin. There is also a **motor homunculus**; other maps support visual and auditory processing, represent the body in space, etc. Each of these maps reflects the amount of cortex devoted to processing information from that part of the body. Rather than proportional to its size, the body parts of the sensory homunculus are proportionate to their *degree of sensitivity* – to the number of sensory receptors in that particular part of the body. As a result, you will feel something touching your hand with higher resolution than your elbow because more of the cortex is dedicated to hands than elbows.

Keeping with the perception of touch, according to the computational theory of mind:

1. The cortex is comprised of columns of cells. Each column represents a "computational module" which performs a particular and specialized function. The number of columns correlates with the size of the cortical area dedicated to that part of the body;
2. Incoming sensory data is directed to the **somatosensory cortex** in the parietal lobe. Sensory input is processed hierarchically, from the **primary somatosensory cortex** which includes the

sensory homunculus, to the **secondary somatosensory cortex** which manages more complex matters;
3. Input is handled in a serial fashion through a progression of computations to discern, decipher, associate, and refine the data;
4. Concurrently, the data is processed in parallel pathways which perform different kinds of computations; and,
5. The information of unimodal processing is then merged with the other kinds of information in multi-modal areas, and integrated such that the final product reflects all the senses in a blended, synthesized picture.

To then *act* on that perception requires the **motor system**. Whereas the sensory system builds from the bottom-up, the motor system starts with a plan of action and breaks it down into a series of steps. The **primary motor cortex** is therefore the *last* stop in the chain – it conveys the minute steps necessary to execute the plan.

THE NETWORK DYNAMICS MODEL

In his book, Networks of the Brain, Dr. Olaf Sporns makes a simple assertion: the reason it's so difficult to assign mental faculties to particular brain regions is because we're paying too much attention to the faculties and their neural signature, and not enough to how the brain actually works. He suggests the answer lies in network dynamics -- what's important is "how brain networks can generate different classes of dynamic behavior and how these dynamics map *onto cognition*" (emphasis added),[6] rather than the other way around. The network dynamics model agrees with the notion of **small world connectivity**, that most neurons are connected by short paths and the densest connections are close by. It agrees that clusters of local circuitry form communities of **nodes**. It proposes that nodes form functionally-specific and often overlapping **modules**, and modules are linked together through **hubs** which allow information to flow between modules and provide system-wide coherence and integration. But it is there that the similarity with other models ends.

The network dynamics model sees cognition as a network phenomenon: cognition and other complex functions are the result of a dynamic, spontaneously active, recurrently connected brain; the "interplay of functional segregation and integration and the continual emergence of dynamic structures that are molded by connectivity and subtly modified by external input and internal state."[7] To understand this, we must set aside the idea of computations and neuron-to-neuron connectivity and think instead of the brain as a dynamic pattern creator. In such a brain[8]:

- Neurons act independently *and* collectively;
- Clusters of neurons are functionally segregated *and* structurally connected, meaning they perform distinct functions and are linked together;
- Functionally segregated areas are integrated by **convergence** and **synchronization** – through the melding (converging) of inputs and the coordinated (synchronized) firing of distant populations of neurons;
- Information processing occurs on multiple scales and throughout the brain. It can be serial but may also reflect "hierarchical modularity" – meaning overlapping modules are "nested" within other modules to conduct higher, multi-modal processing;
- **Feed-forward** and **feedback** loops and **recursive patterns**[9] connect nodes such that higher areas feed information back to lower ones, and vice versa;
- Operations are distributed throughout the brain but rather than occur in parallel, they are pieces of rapidly changing networks that are shaped by the task at hand, the internal state of the body, and the sensory input received; and finally,
- Heterogeneous and homogeneous networks meld together "on the path to global synchrony."[10]

The networks dynamics model is not about functions or regions, wiring diagrams or computations. This is a model of **embodied cognition**:

where the brain is a "dynamic, spontaneously active, and recurrently connected system"[11]; where sensory input "awakens" the cortex, as opposed to directing it to fire in a certain way[12]; where functions are the result of network interactions that form based on the nature of the task, only to then dissolve and rapidly reconfigure as situations and stimuli change; where cognition and other complex mental processes are the result of "the continual emergence of dynamic structures...molded by connectivity and subtly modified by external input and internal state."[13]

The Networks Dynamics Model is still just a theory about how the brain works. And we continue to have the problem of, "Am I my neurons?" Dr. Sporns considers this a problem of **neuro-reductionism**: he says it's correct to suggest that mental states and functions have a "physical basis," but wrong to say we're nothing more than our neurons.[14] Rather, human beings are the sum of information, social cognition, embodiment, and network dynamics.[15] He also suggests that network dynamics might explain how the brain ultimately creates a unified, seamless picture of reality: by merging diverse inputs (convergence), coordinating neural firing across distant groups of neurons (synchrony), and **reentrant dynamics** whereby feedback and feed-forward loops are "concurrently engaged," incoming streams of information are unified and presented in a contextual framework.[16] The net result is the seamless whole we experience moment-to-moment.

A
Closing Thought

A Common Myth about the Brain: There's a little man at the controls.

This is known as "the homunculus problem" and what's important to remember is that just because one has the feeling of being in control does not mean there is a controller. We are schooled from a very early age to attribute actions to the **mind**. We think of ourselves *as* our mind and others as their mind. We learn to associated the mind with a sense of **agency** and experience our mind as "the doer." Over time, "we accumulate the picture of a virtual agent, a mind apparently guiding the action."[17] The problem is that the mind creates the "*experience* of agency," and that experience is "derived from the perception of thoughts and action – *not* as a direct perception of an agent" (emphasis added).[18] So there is no little man at the controls. A much more interesting question is: How do we understand a sense of agency, a feeling of control, and the notion of the mind *absent* a controller?

8

The Human Brain and What Makes Us Unique

Perhaps you've noticed over the past few chapters, the three special properties of the brain. The brain's **embodied** nature is pretty obvious: the brain lives in and through a body. Neural pathways form in the body; the brain-body-environment interaction creates our subjective inner state through which we perceive a seamless, coherent whole. The brain's **emergent** property is also evident: in its self-organizing capacity, **neuroplasticity**, and the way network dynamics are shaped by the task at hand. The brain's **conservative** nature is present as well: in how evolution keeps what works, older brain regions are adapted for new purposes, and the prefrontal cortex prefers a controlled, stress-free environment.

This contrasts with the common belief that the human brain is like a computer. Both the brain and computers process information: they take in data, process it, and convert it into generally useful information. However, there are many ways in which the human brain is *not* like a computer[1]:

1. Computers are fast. Neurons send signals through **action potentials**. It's a biological process; it takes some time.
2. Computers are serial. To maximize efficiency, the brain runs many parallel tracks concurrently to associate as much as possible with whatever is being thought.

3. Computers are generally reliable and assembled according to plans. The accuracy of neural memory, for example, is questionable at best, and as we've seen, genes provide a template not a blueprint.
4. Computers only have so many connections while the nervous system has 100 billion neurons, each with some 10,000 connections.[2]
5. Computers rely on microprocessors. The brain uses neurons; it's been said *each neuron* is like a computer.[3]
6. Computers generally work alone. The brain's work happens in a context, in a body, and often in the company of other brains.

But the computer metaphor is an especially pliable one and has had mass appeal. Most neuroscientists think of the brain as a kind of computer, although they understand the brain works in a "fundamentally different way" than a PC.[4] On the other hand, the idea of a brain-to-computer interface to download entire areas of expertise from one mind to another doesn't work…because each brain uses its own "idiosyncratic" storage format based on past history, genetic, and other factors.[5] In other words, it doesn't work precisely because of how the brain is *not* like a computer.

WHAT SETS THE BRAIN APART

There are at least three sensibilities which set the brain apart from a computer. These include our capacity for *meaning, stress, and knowing* – all three of which are so important they are built right into the brain.

1. <u>Meaning</u>. The information the brain processes is energy with meaning.[6] Because we rely less on instinct than other animals, meaning has special significance for humans; in fact, some say we are on a lifelong "quest for meaning."[7] Meaning increases survival; it's one of the main ways we fight death and keep focused on living.

    Meaning is directly tied to the **viscera** and the **seeking systems**. Through the viscera, we *feel* meaning. Similarly, the seeking systems detect imbalances in the body and motivate us to look for food, find a partner, or meet some other need. While these systems originally focused on the motor system and basic needs, they have since evolved as the source of our higher drives, the "cognitive interests" that bring "existential meaning to our lives."[8] Interestingly, the seeking systems may be related to how the brain forms *causal relationships*.[9] When we are seeking something and two events happen together, we "intrinsically" assume one caused the other. Although these relationships are often incorrect, this tendency to link causal events may contribute to a "consensual understanding of the world" – to an agreement among social group members that the world is this way and not that.[10]

Neural activity that generates meaning depends on signaling in the brain; on neurotransmitters such as glutamate, acetylcholine, GABA (gamma amino butyric acid), as well as melatonin, serotonin, glutamate, and dopamine. The same neurotransmitter can affect different parts of the brain in different ways: **dopamine** enhances the fluidity of movement in the **substantia nigra**; it is related to pleasure and reward via the **ventral tegmental area**, and helps sequence thoughts through the ventral tegmental area and the prefrontal cortex. Signaling also involves **hormones** which are released into the bloodstream and act on distant targets. Too much of the stress hormone **cortisol**, for example, *increases* learning in the amygdala and *decreases* it in the hippocampus. Another term for "stress hormone" is **glucocorticoids**; cortisol is a type of glucocorticoid.

We learn what is meaningful through our social relationships. **Mirror neurons** play a role in social cognition, helping us understand another person's actions, imitate those actions, and simulate them to try to understand the intention behind them. **Spindle cells** have been found in humans, great apes, elephants, and certain dolphins and whales – all social creatures -- and may be related to tying together multiple areas of the cortex to support complex functions.[11] **Maternal attunement** is also known to increase brain connectivity and lay the foundation for self-control, self-esteem, and sociality.

2. <u>Stress</u>. Stress is also built right into the brain. While computer systems can be stressed, computers don't *feel* stress. For humans, stress is part of our ability to feel and grow. It can be triggered by a variety of biological, psychological, and/or social factors. A certain amount of stress is necessary for learning: it's stimulating, and creates motivation and curiosity. In addition, we each have a warning system through the amygdala, the autonomic nervous system, and the **hypothalamic-pituitary-adrenal axis** to detect and respond to a threat. In the face of a threat, the autonomic nervous system engages the **sympathetic nervous system** which begins diverting energy away from thoughts and thinking. It sends that energy to our muscles, speeds up our heart rate, and increases our blood pressure, creating what's called a **threat response**. Once the threat has passed, the **parasympathetic nervous system** takes over, relaxing our muscles and returning us to a more balanced state.

    The threat response becomes a problem when we get stuck in arousal mode without the natural return to a calm, relaxed state. This otherwise normal cycle then becomes **stress**. It is particularly harmful when the stress is chronic: the body is flooded with stress hormones, our subjective inner milieu is in constant turmoil, and our seeking systems become distorted as we look for anything to numb the pain. When the state is prolonged, it can cause the body to deplete nutrients more rapidly, weaken our immune system, and increase fat storage.[12] As Dr. Van der Kolk, author of <u>How the Body Keeps the Score</u>, tells us when we're having an inner battle, it's largely "played out" in the viscera, causing both "physical discomfort and psychological misery."[13] He believes that becoming aware of what we're feeling *on the inside* – by engaging the medial prefrontal cortex and its abundant subcortical connections -- is the critical first step in the healing process.[14]

When we're under stress, it's hard to find meaning in anything. But stress also affects **neuroplasticity**. As we said, too much cortisol increases learning in the amygdala: it increases **long-term potentiation** (LTP) so we remember things that are important to survival (See Chapter 5). *Short-term* stress, such as a moderate challenge, also increases long-term potentiation in the **hippocampus** and facilitates learning. Hyper-arousal, however, has just the opposite affect: we can't learn and if the stress is chronic, it can disrupt neural pathways and kill brain cells. Stress can also overwhelm the prefrontal cortex: because of it abundant connections and shared authority with the amygdala, when we're under stress, the speed of the amygdala makes us more susceptible to re-acting and/or relapsing into bad habits.

3. <u>Knowing</u>. The third capacity that sets the brain apart from a computer is the feeling of knowing. Knowing is a mental *sensation*. We *know* when we've learned something, know what someone means, and know the right thing to do. In his book, <u>On Being Certain</u>, Dr. Burton talks about the *feeling* of knowing and how, like other bodily sensations, it is influenced by "everything from genetic predispositions to perceptual illusions."[15] The feeling of knowing is an extension of our base of knowledge and relates to the brain's need for certainty. Certainty – and the feelings of rightness, conviction, and correctness that go along with it – operates outside reason,[16] and often outside consciousness as well.

Knowing is a uniquely human sensation, we have a *need to know*. It lies at the heart of **social cognition**, our theories about other people's mind, and is intimately linked to our **sense of self**. Knowing is something we feel inside; it's part of our identity. Like stress and meaning, knowing requires a balance. Too much meaning becomes obsessive. Too much stress becomes destructive. Too much knowing and we turn it on ourselves and others. It can lead to lifetime of self-proving and telling others what to know; in effect, we take up too much space in the room.

Meaning, stress, and knowing are built right into the human brain and make us distinctly unlike a computer. All three happen in the body, are emergent in nature, and have a conservative leaning. And they all point to the **wisdom** of the brain: to the fact that **safety is personal**. From a safe place, we can find just the right amount of meaning, stress, and knowing, and tap into the grace, resilience, and understanding that flow from it.

A
Closing Thought

Moments of Clarity in Neuroscience:

A perception is the association of an idea with a stimulus.
A thought is the association of an idea with another idea.

Change involves either inhibiting an existing pattern…or strengthening an alternative one.

Making up your mind may be a process of discovering an idea hidden in your unconscious…

In an average 16 hour day, **we think 4,000 thoughts**.

It takes 20,000 genes, **80% of the entire genome**, to create the brain.

An Emotion is a set of physiological (body and brain) responses when the brain faces a challenging situation.
A Feeling is our conscious awareness and experience of those responses.

Creativity is the ability to restructure our understanding in a non-obvious way.
Insight is an unconscious shift in mental perspective that solves a problem.

Intuition is an inference drawn from past experience; it's a cognitive, rather than a sensory, experience.
Intuition is when something that is known *implicitly* becomes *explicit*.
Intuition is an inner feeling of knowing.

A gut feeling is a feeling of knowing without the awareness of the thought or line of reasoning that caused it.

Certainty is a mental sensation not based on reason.

Section III

How the Brain Thinks

9

Thinking as Representations

In this section we shift from how the brain works to how we think...how we form thoughts and ideas. The brain works like as "associative machine": one idea triggers multiple other, different ideas...in a "spreading cascade" where a single word evokes memories...which trigger emotions, stirs feelings, bring places and faces to mind...and all of this happens so quickly, it yields "a pattern...that is both diverse and integrated, [that is] *associatively coherent.*"[1] The brain forms associations, and from there, thinking flows.

PATTERNS FIRST

Thinking is based on **patterns**. For early humans, the ability to pick out the tiger in a forest of trees was a basic survival skill. In fact, thinking, perceiving, discerning, understanding, feeling, and acting are all based on patterns:

- The brain is both a pattern finder and a pattern maker: it takes in information, associates it with stored knowledge, and creates a pattern. The brain continually retrieves and applies those patterns to future thoughts and situations; ensuring every new thought and idea is built on the past. The human need to find patterns is so strong that we persist in looking for them even when we're told none exist.[2]

- Patterns feed the brain's desire for certainty, making both prediction and predictability possible. Through patterns, the brain can predict regularities and attend to irregularities. This fosters predictability – a feeling of knowing, familiarity, and control. Patterns help keep stress at a minimum, and through them, we're able to derive rules which we can apply across the board. The brain handles most of our experiences through patterns, reserving higher order processing for unique or novel situations.
- Patterns appeal to the brain's need to make sense of its surroundings. Every experience leaves an imprint or trace of energy on the brain. The brain then groups these based on their similarities: patterns form categories, categories become generalizations, and beliefs, assumptions, and expectations flow from there. We use patterns to identify group members and to define "otherness" – those who are outside cultural norms (another word for pattern).
- The brain structures life around patterns: we see patterns in random data, adhere to a daily routine, have pre-programmed movements and behaviors, and grow **mind patterns**. We have a pattern of causal thinking and a tendency to "infuse patterns with meaning, intention, and agency."[3] In his book, The Believing Brain, Michael Shermer suggests "Belief in a supernatural agent... is universal to all cultures."[4] He sees God as the *ultimate* pattern because it explains "everything that happens."[5]

Various parts of the brain work together to find and form patterns: the **basal ganglia,** among other areas, see patterns from one situation to the next; the **amygdala** trains on patterns of danger and help categorize people[6]; and the cortex recreates patterns in order to recall them. Almost from birth, infants recognize the pattern of their mother's face, a human voice, etc. Interestingly, our ability to see patterns varies among individuals: people with dyslexia, for example, are especially good at seeing patterns in a larger framework -- so much so that companies want to hire them to crack computer code.[7]

Finding causal associations, imposing organization on chaos, and trying to make sense of the world clearly helps the brain manage its environment. Even drawing the wrong conclusion has value: it adds to our list of associations crucial for survival and reproduction.[8] Patterns have another advantage as well: they allow us to draw *inferences*. Based on what we see, we're able to infer about what we don't.[9] We draw conclusions, make assumptions, and form judgments about all sorts of things we have not directly observed. We've even found patterns in death: Kubler-Ross' five stages of grief, which describe the stages people pass through when facing a terminal illness, is widely popular and often quoted precisely because it brings comfort to an often unexplainable world.[10]

In reading what follows, remember, whenever mention is made of a representation -- a symbol, memory, image, rule, map, heuristic, or schema -- there's a pattern at the heart of it. Keep in mind as well, the fact that thinking depends so much on representations, inferences, causal relationships, and patterns points to the **wisdom** of the brain – that **so much of life happens in our head.**

THE REPRESENTATIONAL APPROACH TO THINKING

In Chapter 7, we discussed the **computational theory of mind**: how the brain takes in sensory data inputs, performs massive computations on those inputs, links inputs with symbols or representations stored in the brain, and translates the results into thoughts and actions. There, we considered the computational part of this theory. We will now look at the representational piece. Together, they form a theory about how we think.

The representational approach to thinking involves taking in sensory information as *perceptions*; using memory to assemble those perceptions into "*representations*"; translating those representations into *ideas and feelings*; and, turning those ideas and feelings into *action*. It feeds the brain's **conservative** nature because past experience in the form of memories and related associations get linked together to form an idea or thought. Each idea or thought is a set of neural representations dispersed throughout the brain. By linking inputs with representations, the brain is able to transform data into information. So on a hot, sweltering summer day, the following representations and related associations might come to mind: sweating…take a swim…go for ice cream. The idea to go for ice cream is a set of neural representations dispersed throughout the brain, based on past experience, which is quickly assembled to solve the problem: get the keys, go to the car, start the engine, drive to the store, pick the flavor, pay for the ice cream, and grab a spoon!

REPRESENTATIONAL THINKING AND THE VISION SYSTEM

Nearly half of the cerebral cortex is devoted to vision and visual processing, and while it's often thought to work like a camera, that's really not the case. Visual stimuli come in through **sensory receptors** in the eyes, to the **thalamus**, and on to the **primary visual cortex** where it progresses through three levels of analysis. The information is then transmitted through two separate but interacting streams[11]: the **ventral stream** – also known as the **"what"** or more recently, the **"how" pathway** -- flows through the **temporal lobe**[12] and stores information about shapes, sizes, faces, etc.; the **dorsal stream** -- or **"where" pathway** -- flows through the **parietal lobe** and helps integrate vision with motor action.[13]

As opposed to a camera, the brain breaks down a scene into component parts, applies a set of known rules about how the world works, and then literally "guesses" based on past experience.[14] The brain:

- Separates the foreground from the background, and figures or objects from the background;
- Compares it to internal representations, matching visual input with memories of shapes and other stored associations; and,
- Then assembles this distributed process into a unified presentation.

Thus the entire process is *constructive* -- it builds from the bottom-up: our memories, rules, and expectations get incorporated into the properties of the stimuli and shape what we ultimately see.[15] Interestingly, the visual signal in the thalamus incorporates what's known as a "gating function" which is linked to attention: working with other parts of the brain, the gating function "modulates" (or adjusts) how strongly a response is to a given stimulus based on how important that stimulus is at this particular moment.[16] This happens in the transmission of auditory (hearing) signals as well. As we've said, perception is not a direct transmission. It's constructed.

REPRESENTATIONAL THINKING AND TAKING ACTION

Once sensory information is processed, action often follows. The motor system operates from the top-down: it starts with the plan and breaks it down into a series of steps. Different types of movements require different levels of control: **reflexes** can be carried out largely by the spinal cord; heavily-rehearsed, **procedural** and other learned sequences also require little cortical input; **goal-directed** movements, especially new or adaptive ones, require considerable and high level cortical control. The motor system is a distributed process and is organized hierarchically: in the brain's frontal regions, information flows from the **prefrontal cortex (PFC)**, to the **premotor cortex**, and onto the **primary motor cortex**, interacting with the basal ganglia along the way.[17] The prefrontal cortex is thought to "represent and coordinate complex goal-directed" behavioral sequences; the premotor cortex handles movements "defined by goal and trajectory"; and the primary motor cortex represents and organizes movements "defined by the muscle groups that produce them."[18]

The prefrontal cortex plays a coordinating role in all goal-directed activity; on a global level, and all the way down to telling neurons in the primary motor cortex when to fire and in what sequence. The prefrontal cortex can be thought of as a repository of goal-related "hierarchies" – sets of plans for accomplishing given tasks. At this level, perception (sensory) and action (motor) are wholly intermingled: one actually shapes the other. Moreover, at the highest levels, neurons encode not only the physical features of the stimuli (the "what") and the force and direction of movement (the "where"), but also the relationship between the body and the object *relative to the goal*.[19] So if you're thinking about hitting a baseball, incoming sensory information (the ball, its velocity, its trajectory, etc.) shapes the motor output (how you're going to hit the ball) towards the goal (of having it sail out of the ballpark).

To assemble a goal hierarchy, the prefrontal cortex draws representations from long-term memory and ties them together on a temporary basis in **working memory**. General associations are fed into working memory from the interactions between the **lateral prefrontal cortex** and the **hippocampus**, while the **orbitofrontal cortex** and the **amygdala** provide input on risks and benefits[20] and add feelings and affective values to the mix. It's interesting to note that in assembling an action plan or solution, the brain tends to think narrowly about how an object can or should be used.[21] This is one of the reasons why creativity is so difficult.

The **premotor, supplementary motor** and **primary motor** cortices encode hierarchies of representations of movements. Once a plan is formulated, the primary motor cortex is the last stop in the processing chain and translates the goal into a specific sequence of muscle movements with the aid of spinal cord circuits. All of these regions work with the frontal lobe attention network which focuses attention and allocates mental resources; the **cerebellum** which monitors errors and helps coordinate movement; the **anterior cingulate cortex** which compares execution with intended results; and the **basal ganglia** which helps initiate movement and assists if conditions change or the plan isn't working. **Feedback** at every step, via **reentrant** (also known as recurrent) **networks,** continuously updates and refines movements, and/or adapts to changing conditions. Once executed, the memory library of hierarchical representations is then updated to incorporate recent events.

A
Closing Thought

A Neuroscience Question: How does the brain give rise to the mind?

How does something physical – like the brain, create something non-physical – like the mind? In popular literature, this is referred to as the **"mind-body problem"** and there are several schools of thought. Dualists believe the mind is separate from the body and made of distinct substances. Monists hold that the mind and matter are made of one and the same substance, although its specific nature is not expressly stated.[22]

The debate over the physical brain and the non-physical mind has run throughout the history of neuroscience. It's referred to as the "**hard problem.**" It contrasts with the "**easy problem**" which involves identifying the neural correlates of mental functions, including those of **consciousness**. The hard problem has subsequently been reframed to ask: "How does matter create consciousness?" It's said most neuroscientists are material monists: they believe the mind and brain are reducible to the same substance and that substance is physical.[23]

Steven Pinker suggests that the computational theory of mind *solves* the hard problem. Referring to it as "one of the great ideas in intellectual history," he argues that under this theory, "beliefs and desires are information, incarnated as configurations of symbols" (i.e. representations) which are "the physical states of bits of matter."[24] In others words, patterns of neural connections symbolize things in the world, forming thoughts, actions, etc. out of matter. If he's right, then the answer to the question, "Am I my neurons?" is a resounding yes.

But not everyone agrees. Some say reducing human beings to neurons doesn't work because of the property of **emergence**: the mind and consciousness emerge from the brain similar to how flight emerges from an airplane.[25] Other cultures approach this question from an altogether different vantage point. In the eastern Sankya System, the entire universe is a manifestation of consciousness that gradually takes form (or "encases itself") in matter.[26] That means that consciousness comes *first* and is infused *into* the body. Consciousness uses the body to experience itself. What do you think?

10

Thinking as a Cycle

In the **perception-action cycle**, the entire cortex is devoted using representations to generate *action*. This model of thinking differs from the representational approach in several key respects[1]:

- First, information flows from the back of the brain to the front: perception (sensory) is built from the bottom-up in the back or **posterior** part of the brain, while action (motor) occurs top-down in the front or **anterior** part of the brain;
- Second, the flow forms a cycle: sensory information feeds into and shapes motor behavior and vice versa. Sensory and motor systems interact with the central nervous system and internal body milieu, the emotional substrate, the correction-monitoring parts of the brain, and the environment; and,
- Third, a very different role is envisioned for the prefrontal cortex.

THE PERCEPTION-ACTION CYCLE APPROACH TO THINKING

The perception-action cycle[2] divides the brain into a posterior or "sensing section" and an anterior or "doing section." The model assumes a network or widely distributed view of the brain where connections are formed through **associations** from learning and experience. It also includes **cortical and subcortical loops** that tie together the orbitofrontal region, limbic system, and dopamine systems; the latter of which serve to motivate behavior.

In this model, the brain performs five principle "cognitive functions": attention, perception, memory, intelligence, and language. The prefrontal cortex (PFC) performs sub-tasks of these functions, including executive attention, planning, and decision-making; each is accomplished through a network and not by the prefrontal cortex alone. While the prefrontal cortex in the representational approach functioned much like a command center -- coordinating motor activity, maintaining goal hierarchies, sequencing neural firings, etc. -- in the perception-action cycle, the prefrontal cortex plays a more integrative role, enabling "sensory and other inputs [to] determine and guide commands and decisions."[3] The prefrontal cortex is seen as working under the "continuous constraints" of long-term memory, instinctive and emotional influences, the limitations of sensory input and motor output, and from ongoing feedback, such that it "cannot generate, let alone sustain, any action, but it can orchestrate them all."[4]

The perception-action cycle is the means by which goal-directed activity is organized and executed in the brain. Sensory input ascends from sensory receptors to the **posterior cortex** where it is processed from lower to higher levels and from unimodal to multi-modal association levels. Association levels contain **perceptual memory**, a bank full of representations of images from, and conventions about, the external world which are used to form perceptions. Some percepts include representations of actions which alert the frontal regions to prepare to take action.[5] Working memory, supported by the dorsolateral prefrontal cortex, then fills in the "temporal gaps between perception and action"[6] through **reentrant loops**. These loops tie together the representational networks of the anterior and posterior parts of the brain and thereby avoid the need for a central executive or command center.

The prefrontal cortex includes **executive memory**, a future-looking memory, through which action plans are formulated.[7] The prefrontal cortex also integrates emotional inputs, drives, and affective values from the **orbitofrontal cortex**, and other subcortical areas. While the prefrontal cortex takes part in all decisions and actions, the dorsolateral prefrontal cortex (DLPFC) focuses more on goals, planning, and related executive functions, while the orbitofrontal cortex (OFC) is especially involved in emotional cognition and bodily state awareness. Action plans flow from the prefrontal cortex -- the "highest cortical level in the motor hierarchy" -- to the **premotor cortex** and on to the **primary motor** cortex, becoming increasingly solidified and detailed in preparation for execution.

In the perception-action cycle, thinking is a continuous flow of information -- from sensory to motor, posterior to anterior, subcortical to cortical, and between the organism and the environment. In effect, sensory input from the posterior brain, thoughts and meaning from working memory and the dorsolateral prefrontal cortex, emotion and value from the orbitofrontal cortex -- all get passed forward to the anterior prefrontal cortex to fine tune, integrate, and translate into action. The model includes continuous feedback from the environment and the central nervous system to gauge effects, correct actions, and ensure goals are met. And while representations play an important role in the transforming sensory input into a plan, they are subservient to the overriding goal of enabling the organism to act.

A SUBCORTICAL APPROACH TO THINKING

Authors Koziol and Budding emphasize the role *subcortical structures* play in the thinking process. In their book, Subcortical Structures and Cognition, they contend:

- The brain has two distinct systems for knowing what to do that have fused together over time: a stimulus-driven system for routine and survival-based behaviors, and a problem-solving system for learning new behaviors and resolving novel problems[8];
- Cognition and cognitive control may be an "evolutionary extension" of the motor control system[9]; and,
- All problem-solving behavior involves "breaking a problem down into stimulus-based characteristics."[10]

Against this backdrop, they point to two critical subcortical systems heavily involved in thinking: the **cortico-basal ganglia/striatum system** and the **cerebro-cerebellar system**. Both systems form a loop in the brain: they originate in the cortex, pass through multiple subcortical regions, and then reenter the cortex, terminating near where they begin. In addition, their *cortical inputs* are always excitatory while their *subcortical outputs* are largely inhibitory, meaning the subcortical areas help decide whether or not information is returned to the cortex.[11]

In the cortico-basal ganglia/striatum system, the **basal ganglia**, via the thalamus, serve as the interface between the cortical and subcortical regions. The basal ganglia/striatum circuit specializes in **pattern recognition** and categorization, and conveys information to the cortex whenever novel, non-patterned situations are present. It is extremely sensitive to context and especially rewards, and is able to learn both desired behaviors and those to be avoided. In addition to initiating movement, the basal ganglia participate in deriving the *intention* behind movement and, alongside the cortex, update working memory.[12] In sum, the basal ganglia and the cortex "interact continuously" in most activities and in many instances, the basal ganglia/striatum circuit "biases" the prefrontal cortex to act or to act in a specific manner.[13]

In the cerebro-cerebellar system, the **cerebellum** does more than learn and coordinate compound motor movements. It receives input from the senses, as well as information about goals and intentions. It then aggregates this information into a "unified pattern" of action and creates a "map" against which to compare "the course and outcome of the behavior."[14] The cerebellum signals the cortex to release the behavior, while maintaining the map for comparative purposes. The authors suggest the role of the cerebellum should be thought of as controlling the "*quality* of behavioral output" (emphasis added).[15]

Koziol and Budding's message is clear: when we think about thinking, we need to look beyond the cortex. In their view, "the cortex is actually driven by basal ganglia and cerebellar input, and in this manner learns to perform a behavior more quickly, more accurately, and more automatically…"[16]

SOME ADDITIONAL THOUGHTS ON THINKING
Irrespective of the model, several points deserve clarification:

1. Sensory, cognitive, and motor processes do not occur sequentially. It may *feel* like we perceive things, think about them, and then act, but sensory, cognition, and motor processes all act together and shape one another.[17] In many instances, sensory and motor areas process information simultaneously.
2. Emotions cannot be isolated from decision-making, intelligence, behavior, etc. Emotional input is processed in the brain concurrent with other data, and absent emotions, nothing can be perceived, no decision can be made, and no action can be taken.
3. While the dorsolateral prefrontal cortex (DLPFC) focuses more on goals, planning, and related executive functions (sometimes misleadingly called "rational factors") and the orbitofrontal cortex (OFC) is especially involved in emotional cognition and bodily state awareness (sometimes misleadingly called "emotional factors"), these regions should not be confused with the two hemispheres of the brain. They are present in *each* hemisphere.
4. Most of the brain's structural connections do not process input or output, but are *connective* in nature: they connect nodes within the brain in recurrent patterns so information can be passed on, massaged, and interpreted.
5. It's important for helping professionals to recall, there is a difference between **executive control** and **executive functions**: psychologists talk about executive control, neuroscientists speak of executive functions. Executive control is something we exert over habitual or unwanted tendencies. Executive functions are often associated with intelligence.

6. The prefrontal cortex, with its abundant connections, can overrule impulses, habits, instincts, etc., but it takes effort. It's equally true that a "*primitive* neural system can easily overmaster an immature or poorly developed *advanced* neural system" (emphasis added).[18] This is an important point for helping professionals to keep in mind: just because we have higher processing capabilities doesn't mean we use them; and just because we use them, doesn't mean they're well developed or honed.

Recommended Reading
Thinking, Fast and Slow by Daniel Kahneman
Neuroscience for Clinicians or The Dao of Neuroscience by Simpkins and Simpkins

A
Closing Thought

A Neuroscience Question: What is the relationship between the brain and the mind?

We know we learn to think of ourselves as our mind and others as their mind; that we link the mind with a sense of agency when, in fact, the mind creates the *experience* of agency. We also know that everything in the mind has been put there; perhaps, in some fashion, in the brain as well. Consider the following two lists:

The Brain is...	**The Mind is...**
The exchange of energy	The exchange of energy
Manager of the inner/ outer environment	Layers of inner process
A pattern maker and finder	The flow of information
A prediction maker	A meaning maker
An analogy maker	What the brain does
Physical	Psychological
The organizer of reality	The brain in action
An organ of adaptation	Where the brain ends
An hypothesis maker	The explainer, explorer
An information processor	The embodied brain
For prediction and control	Intelligent action
Inhibitory	Subjective awareness
A belief or experience designer	First person perspective
The source of causality	A feeling of resonance
Life's manager	The same as the brain

The **wisdom** of the brain may offer a solution to this question. Perhaps the general purpose of the brain is to manage our internal and external environment. Perhaps the mind is where "the self" lives. And together, the brain and mind seek three "goals": **to make sense of the world, to maintain a coherent sense of self, and to sustain some semblance of control at all costs**. What do you think?

Section IV

The Brain and Individuality -- Parenting, Emotion, and Past

11

Parenting and How the Brain Develops

Every client who comes to us experiences their own brain-body-environment interaction. They have their own neural pathways that form and dissolve; their own ideas about meaning, stress, and knowing; their own capacity for change and self-management. We don't know what pain feels like to them, how safe they feel, or how they feel safe. Each comes as a unique individual and the differences run deep…and it all starts with parenting. Parenting builds brains. Parents shape our biochemistry and mold our safety-danger continuum. They sculpt our psychological nature and instill in us self-esteem and self-control. They are the conveyors of culture and teach us about fairness and trust. The **wisdom** of the brain is that **individuals are built through relationships**, especially with those who dominate our lives.

To learn language, how to read faces, and appropriate social behavior, infants and children need other people and in that way, we are innately and necessarily **social**. Human babies are born prematurely compared to other mammals – many of the cognitive and motor skills chimpanzees are born with, a human infant has to learn during their first year.[1] They have an exceptionally long period of post-natal development during which time they are wholly dependent on their caregivers. In fact, nearly 70% of the structure of a human infant's brain is developed post-birth,[2] which means that the bulk of the brain's connections are **experience-dependent**. And yet, equally important is the **initial genetically programmed connectivity** which occurs in the womb. As a result, pregnancy, maternal attunement, and caregiver attachment are *the* key determinants of our future well-being.

EVOLUTION AND NURTURANCE

Somewhere along the evolutionary path, it occurred that nurtured infants survive better. Nurturance is thought to have its evolutionary roots in the neurochemistry that controlled mating and egg-laying in reptiles.[3] These neurochemical systems lie deep in the subcortical regions of the brain and promote sexual, maternal, and distinctly social behavior in mammals. According to animal research, neuropeptides such as **oxytocin** and opioids such as **endorphins** are activated by "pleasurable pro-social activities" such as grooming, play, and mating.[4] Along with **prolactin**, they are thought to foster sexual behavior, social bonding, and nurturance, and the social feelings of warmth, love, connection, and safety.[5] The female brain in particular is primed to care for her young. In fact, the evolution of the maternal circuit may have "initially led creatures down the mammalian path."[6]

Dr. Jaak Panksepp, who identified the four major "emotional operating systems" (See Chapter 3), calls this nurturing capacity the "care circuits." He suggests these circuits evolved alongside the Panic System, the neural networks that prompt mammals to express and respond to distress cries, especially from infants. The evolutionary roots of the Panic System may date back to "place attachments in reptiles" and the "affective mechanisms for pain and...thermoregulation."[7] That means that the circuitry which drives mammals to respond to the distress cries of infants -- the *psychic* pain of separation -- derived from the mechanisms related to the experience of *physical* pain and warmth.

These neurochemical shifts were accompanied by certain neural and anatomical changes such as a specialized middle ear for listening, the ability to make facial expressions, and the capacity to "signal and detect vocalizations of pain, distress, and joy,"[8] leading to the evolution of a more *social* nervous system. The increasingly social nature of the brain that resulted includes the **cingulate cortex** which is thought to make **resonance**

behaviors possible[9] – those behaviors we exhibit when we are especially attuned or in sync with another person; the **anterior cingulate cortex** which participates in the experience of both physical and social pain; and the formation of **spindle cells** which develop after birth and are believed to support complex functions like attention and a self-concept.[10]

THE DEVELOPMENT OF THE BRAIN

Why is pregnancy such an important time in the development of a human brain? It's because:

- The brain begins developing the 18th day after *conception* and neural patterns begin forming by the *24th day*[11];
- During the cell proliferation phase, an estimated 250,000 cells are produced *every minute*[12];
- **Corticogenesis** -- the production, typing, and placement of neurons in the cerebral cortex – begins during the *first quarter* of gestation[13]; and,
- There are more neurons in an infant's brain than there are connections on the Internet; as many as there are stars in the Milky Way. Almost all of them are generated in the womb – they form *prenatally*.[14]

Much of the nervous system of a newborn is like a template, reflecting the hardwired circuits of a basic plan. In their **template role**, genes direct the organization of the nervous system, guiding the development and migration of neurons to their destinations in the brain and body. Once an egg is fertilized and cell division begins, the developing group of cells forms three main cell "lines": the **endoderm** will become the digestive and respiratory systems; the **mesoderm** will form into muscles, bones, and connective tissue; and from the **ectoderm**, the skin, brain, and nervous system will develop. During development, the ectoderm forms the neural plate which later becomes the neural tube and the spinal cord; a cavity forms to hold the primitive brain, while the cerebral cortex covers over the subcortical areas and the brainstem; and cortical enlargement and folding continues for some time thereafter.[15] All of this is "tightly orchestrated" by the expression of genes in the ectoderm telling other cells what to do.

The general stages of brain development from conception to early infancy include[16]:

1. **Cell proliferation** and the production of neurons (neurogenesis);
2. **Cell migration** to their proper location in the brain;
3. **Cell differentiation** into a particular cell type;
4. **Synaptogenesis** or the creation of connections (synapses) among neurons;
5. **Cell death and stabilization**; and,
6. **Synaptic realignment** which occurs during development, during so-called **critical periods**, and throughout life from experience and learning.

Here are some interesting points about the various stages:

- Initially, all cells are stem cells. Two weeks into the proliferation phase, some of these cells lose the capacity to become any type of cell. Their subsequent divisions result in various types of neurons.[17]
- During the process of corticogenesis, neurons migrate based on molecular signals along paths formed by **glial cells**. At this critical juncture, anything that negatively affects the genesis or migration of cortical neurons – such as maternal alcohol consumption, environmental toxins, infections, etc. -- will lead to an ill-formed cortex.[18]
- Somewhere around the 27th week of pregnancy, **synapses** begin to form and reach peak density during the first fifteen months of life.[19] Connection details are added to this initial genetically programmed basic plan both inside and outside the womb – from the sounds of our mother's voice and her movements, to the far

- more diverse postnatal experiences, all of which shape synaptic connections.
- To form synapses, axon terminals "choose" among many possible competing "synaptic partners"[20] using a **growth cone** that moves the axon along and instructs its direction.[21] Once an axon terminal reaches its targeted cell, the contact membrane areas differentiate into pre- and post-synaptic structures.
- Initially, far more synapses form than are necessary. The fetal brain undergoes a process of wholesale "**pruning**" to fine-tune the patterns of connectivity with input from the environment. Neurons left with too few synapses die off, while others begin to stabilize.[22] This process whereby an "overabundance" of synapses is pruned back to a more selective group is known as **plasticity.** It occurs extensively during early brain development. Plasticity also continues throughout life – although on a much more modest scale – because it's how experience and learning are incorporated into the brain.[23]

Once the nervous system starts to work, the experience of the nervous system *itself* begins to shape the neural circuitry. In other words, at some point the initial genetically programmed connectivity of the brain is complete and experience takes over. The details of our neural connections depend our caregivers and the environment; on the amount and diversity of sensory stimulation, the patterning of movements, learning, emotional memories, etc.[24] Rather than neurons dying off, **experience-based reorganization** involves a change in the strength of synaptic connections (**synaptic plasticity**) and **synaptogenesis** (See Chapter 5). Over time these patterns stabilize into what become our neural networks.

12

Parenting and the Brain

Parents build brains: they literally sculpt the neural pathways of communication in a baby's brain. Experience -- and especially *early* experience -- modifies those connections, thereby shaping our individual nature. It is "precisely because" early development is a time of so much "neural growth and organization," that the importance of "early interpersonal experiences" is crucial for how we function later in life.[1]

READY TO LEARN AND NURTURE

Prior to birth, a baby shares the same biochemical environment with its mother: it eats what she eats; absorbs what she inhales; experiences her stress, anxiety, love, and anticipation. By the third month of *gestation*, a baby can grasp; by the fourth, she can suck; at six months, her walking reflexes engage; and at seven months' gestation, she can cry.[2] During the last months of pregnancy, mothers are said to become completely preoccupied with their baby, a condition which continues into the first year of life.[3] When a baby is born, she is utterly dependent but ready to learn; her very survival depends on the caregivers' ability to detect and respond to her needs. To this end, evolution has insured mother and infant are primed to engage, attach, and delight in one another:

- Babies are born with more than 20 **involuntary reflexes** designed to attract the attention of their caregiver[4];
- A baby can recognize familiar faces within hours of birth and will grasp, reach, and turn its head towards its mother;

- A baby learns to follow her mother's gaze and engage in joint attention. This triggers maternal instincts and helps build mother-child bonds;
- Babies learn to read facial expressions and, based on how their caregivers respond to their cries, associate certain expressions with adult responsiveness; and,
- Babies mimic their caregivers, and learn very quickly to equate warmth, safety, dry diapers, and food with pleasure.

All of this means a mother's initial attunement with her baby provides the safety, warmth, and comfort necessary for a newborn brain to develop.

When a baby is born, the **brainstem** is fully formed: a baby can perform all the autonomic life-sustaining activities as well as eat, sleep, and wake; feel hot, cold, hunger, and pain; urinate, defecate, and related functions. With the brainstem largely in charge, most of a baby's early behavior is both **reflexive** and involuntary. At birth, the limbic system "takes off." The **amygdala** is fully mature by the 8th month of gestation, which means the baby is born able to associate fear with a stimulus. By comparison, areas of the **hippocampus**, also part of the limbic system, continue to develop through a baby's first year.[5] The limbic system is said to be shaped by three factors: inborn temperament, genetic makeup, and experience. As a result, "whatever happens to a baby contributes to the emotional and perceptual maps...that its developing brain creates."[6] If the baby feels safe and loved, then the competitive process to form synapses and create neural networks will "specialize in exploration, play, and cooperation"; if the baby feels afraid or unloved, those same neural networks will specialize in ways to manage "feelings of fear and abandonment."[7]

The **cortex** of a newborn is immature at birth and continues to develop well into adulthood. Attention, working memory, executive memory, and planning functions are all thought to achieve "a relative plateau of maturity" by the age of twelve, while higher cognitive functions continue developing into the *third decade* of life.[8] The cortex, and its abundant connections, is shaped by many factors, ranging from parental attention and nurturance to nutrition, language, culture, education, etc. Shaping of this nature occurs through genetic transcription whereby experience triggers genes to produce proteins which in turn strengthen connections and cause new dendrites to grow. Research on rats has shown that an **enriched environment** produces more neurons, more synaptic connections, and more complex dendritic branching.[9] For human beings, an enriched environment means the more challenges, education, diverse experiences, and skills we undertake, the more neurons and synaptic connections form in the brain. And this continues throughout life: "Complex environments provide a variety of stimuli, choices and opportunities, which in turn exercise and sustain mental function."[10]

PARENTING BUILDS *BIOLOGY*

Parental nurturing, and especially the early attachment between mother and baby, shapes a baby's brain *for life*. Obviously, genetics and temperament also play a role, but parental influence is pervasive...and largely invisible. In his books, Dr. Louis Cozolino talks about how parental and especially maternal nurturance becomes **biology**:

1. Parenting builds the initial pathways of communication that form the neural scaffolding for associations, representations, causal relationships, and core memories. With the brainstem in charge, much of an infant's early behavior is based on trial-and-error. How caregivers respond to these behaviors teaches the baby what works and what doesn't, forming early patterns of connection in the brain. These experimental movements become the foundation of later, more organized networks for intentional behavior.[11]
2. Maternal nurturance, attention, exciting stimulation, and arousal stimulate a baby's biochemical systems, including the production of **dopamine, oxytocin, prolactin, and endorphins**.[12] These biochemical systems cause new neurons to grow and new synapses to form; endorphins, in particular, shape initial preferences.[13] These systems also help regulate energy levels at a time when the brain

needs lots of energy to grow, and facilitate the maturation of both the limbic system and the cortex.[14]

3. By creating a consistently safe and soothing environment, parenting increases the growth of **glucocorticoid receptors** in the amygdala, hippocampus, and parts of the **hypothalamic-pituitary-adrenal** axis (HPA) which, in turn, "diminish both the experience of stress and its negative impact on the body."[15] A supportive environment also imprints what's called the "**internalized mother**" on the baby's **unconscious**. This "network of visceral, somatic and emotional memories" reflects our earliest experiences with our mothers, and becomes the basis of our ability to self-soothe and establish solid adult relationships.[16]

4. Parenting molds a baby's safety-danger continuum and in so doing, shapes a baby's emotional nature. Since the older parts of the brain -- such as the amygdala, cingulate cortex, insula, and other emotional processing subcortical areas -- tend to develop faster and earlier, babies have an early **learning bias towards fear**. Fear is a dominant force in human nature; it is easily generalizable and can be associated with nearly any stimulus. In fact, fear may be a baby's first emotion.[17] Most early emotional learning is **implicit**, meaning it happens outside conscious awareness: we don't recall having learned it nor are we aware of its effects over a lifetime.

5. Parenting also forms a baby's **psychobiological regulatory networks and systems** including the hypothalamic-pituitary-adrenal (HPA) axis and the early, top-down cortical-limbic circuitry. The extent and nature of the connections between the **orbitofrontal cortex** (OFC) -- and the **orbitomedial prefrontal cortex** (OMPFC) in particular[18] -- and the amygdala determine our capacity for self-regulation and emotional control. These connections are **reciprocal** in nature such that the cortex can overrule the amygdala *and vice versa*. They are shaped by early experiences as a baby accumulates a "learned history" of what's safe and what's dangerous.

6. Parenting lays the neural groundwork for the especially social nature of the brain. They demonstrate what it means to be in a relationship with another human being, model social skills and appropriate behavior, and train children to read faces and body language, use language to express their feelings, and explore who they are.

Equally important, parents give us our first sense of self. The self begins as a separate "me," a framework for the future self. Both "me" and the self are built through relationships; with our parents, extended family, friends, and significant others. The experience of having or being a self is thought to develop towards the end of the second year of life.[19] The **parietal lobes** give us our sense of self in physical space. It's believed that the **insula** and **anterior cingulate cortex** are involved in "the early development and ongoing organization" of the self,[20] while autobiographical memory adds the narratives that shape our identity. The self is a life organizing principle; once established, "I" becomes the force behind action, an entity to be protected -- the one in control, equipped with meaning and purpose, a reason for living. As part of the early self, parents provide our first experience of *knowing*. Knowing begins as we accumulate a knowledge base, starting with our earliest experiences. As we age, the sensation of knowing can serve us as a vehicle of change and growth, or something we need to assert and defend.

In sum, the parents' response to a baby's needs gives pleasure, comfort, and security that grow into associations, ideas, and thoughts. As Dr. Cozolino states, when consciousness finally comes on board, we emerge "preprogrammed by unconsciously organized hidden layers of neural processing" that form the "core structures" of who we are, how we act, and what we experience moment-to-moment.[21]

CRITICAL WINDOWS AND CONCEPTUAL DOORS

Babies undergo a series of **critical periods** during which time learning occurs at a heightened pace (sometimes called "prepared learning"). The critical period for learning emotional control is from 2 to 30 months.[22]

There is also a critical period for language, for aspects of the vision system and motor skills, for sensory maps, etc. During these periods, neural pathways "wait" for specific information from life experiences to happen in a timely manner. When the information is forthcoming, it triggers the genetic transcription necessary for the brain to continue to develop normally.

During critical periods, the brain is especially plastic. This is because it is still undergoing **synaptogenesis** and subsequent pruning. The learning that occurs during critical periods enables the brain to adapt to its environment, as the synapses of pathways that are used most often stabilize and become more permanent while others die off. That means the brain adapts to a *particular* environment; it is literally reflected in the circuitry of the brain. Because these changes can last a lifetime, it's not uncommon to spend an entire adulthood unconsciously trying to recreate the environment one was wired for as a child[23] -- it's one of the principle reasons for psychotherapy. Moreover, once a critical period has passed, the wiring is essentially set: what a child learns during a given period is difficult to change and will "strongly influence" what the child learns after that window closes.[24] While a child can attain the necessary skills *after* a period passes, early learning is always easiest.

The breadth and complexity of what a child learns from experience is mind-boggling. Babies are born with neural programs for reasoning about objects, numbers, the motivations of others, and physical causality.[25] They are born with a preference for human faces, and animate over inanimate objects. They communicate long before they can talk, taking clues from eye gaze and facial expressions.[26] Language is thought to begin with a need or desire,[27] and babies quickly learn that people and things have labels. By 3 years of age, a child can understand the difference between intentionally doing something and doing so accidentally,[28] and young toddlers know things are not always as they appear, and actions and words do not always match.[29]

The **wisdom** of the brain ensures that **at the core of all of this learning is how to be social** – learning how to interact with others. Very early on, a baby must tackle such complex ideas as:

- There is "me" and "not me";
- Things and thoughts are different;
- There is a real world "out there" and a mental one;
- Thoughts occur in the mental world, in the mind; and,
- People have separate minds and hence, different thoughts.

This is the dawning of "*otherness,*" the knowledge that others are separate and must therefore be understood. It's the basis of early Theory of Mind, the skills we use to discern what another person is doing or thinking. How we grow and develop this fundamental skill set largely accounts for our individuality. For example, it's said children learn fairness in at least two ways: by coming to recognize the negative consequences of an action, and hence understanding that as members of a society, they must follow the rules; or, by learning self-love, concern for others, and appreciating that the needs of others matter just as much as our own.[30] Experiences such as these shape us as individuals. And once the circuitry is established, it's hard to change.

A NOTEWORTHY POINT

Before proceeding, while much of the research focuses on the importance of *maternal* nurturance, any caregiver can give love – aunts, uncles, siblings, grandparents, and especially dads make a baby's world safe, stimulating, warm, and comforting. Moreover, good parenting is not perfect parenting. We're talking about Winnicott's notion of "good enough" parenting. Nevertheless, good enough parenting requires a high degree of selflessness and a willingness to inhibit our own competitive and aggressive impulses.[31]

Recommended Reading
The Neuroscience of Human Relationships, by Louis Cozolino, PhD.
The Neuroscience of Psychotherapy, by Louis Cozolino, PhD.

13

How Parenting Builds the Brain

As helping professionals, the people we work with are products of their early parenting...as are we. Attachment theory is the answer to the question, "How do parents build neural architecture?" The first several years of life is a critical time for synaptogenesis, pruning, and the stabilization of neural pathways: the first 18 months is an especially sensitive time for emotional development, while the second year shifts emphasis from the orbitofrontal cortex (OFC), the orbitomedial prefrontal cortex (OMPFC), and subcortical regions, to the more **executive functions** of the dorsolateral prefrontal cortex (DLPFC) and other association areas. This sculpting happens at a time when:

- The neural architecture is ripe for learning, creating networks that will last a lifetime;
- There are few entrenched patterns of connectivity to overcome or unlearn;
- The limbic system is in full swing, creating the perceptual, emotional, and social maps of the world we will come to rely on;
- The prefrontal cortex (PFC) is not mature, and we are therefore unable to reflect on what's being learned;
- The hippocampus is not online so we can't remember why or when *or that* we've learned something; and,
- We have no stable sense of self...no words to question with and no capacity to choose.

We emerge, as Dr. Cozolino said, "preprogrammed" by unconscious neural processes that form our "core structures."[1] This preprogramming is accomplished through **attachment**; between mother and child, father and child, caregiver and child.

THE NATURE OF ATTACHMENT

Early bonds with "**attachment figures**" form our capacity for self-control, the basis of our self-esteem, and our ability to form lasting relationships with others. The literature on attachment theory dates back to the 1950's, to Harlow and Woolsey and their rhesus monkeys clutching to "wire mothers," and Bowlby and Ainsworth and their work with orphaned children. Both teams found that infants need a secure base: a few select people to go to and connect with, where they can consistently find love and safety. The **attachment bond** is all about feeling safe; it involves:

- The desire to be near the attachment figure;
- Resisting potential separation and feeling distress when separated from that person; and,
- Finding a "safe haven" by seeking out that person in times of fear, stress, or threat.[2]

In effect, a baby needs a safe place to move from; the opportunity to explore, feel challenged, and get excited or aroused; followed by a return to the caregiver where she finds soothing, safety, and security...so she can do it again. Dr. Cozolino refers to this as establishing a pattern of regulation, dis-regulation, and then re-regulation.[3]

Attachment relationships are particularly important during infancy and up through a child's early school years.[4] They give us "our internal working model of security." What matters is not perfection but the parent's *repeated intention* to connect with the child[5] -- to be there with enough consistency that it becomes a **pattern**, a neural pathway in the child's brain. Attachment relationships shape our mental processes, including how we appraise and respond to fear, our eagerness to try and learn new things, how we feel about ourselves and our capabilities, and how well we relate to others. They become the basis for how we make sense of the world.

When a child feels under threat, two circuits in the brain activate simultaneously, although they function in direct opposition to one another: one, in the brainstem, tells the child to flee; the other, the attachment circuit, tells the child to seek out the attachment figure to feel safe.[6] Using what's called the **Infant Strange Situation test**, researchers have been able to measure how children react when separated from their mothers (i.e., is left in a strange situation) and how they react when their mothers return, to categorize the nature of the attachment bond. This research has been replicated many times with the same results.[7] Reactions fall into four general categories: **secure attachment, avoidant attachment, ambivalent attachment, and disorganized attachment**[8]:

1. The securely attached child misses the parent but settles down after the mother leaves and quickly becomes absorbed in play. Caregiving of this type is described as "organized": the child is able to explore with confidence, actively greets the mother when she returns, and moves toward the mother to find safety when necessary;
2. In the avoidant category, the child exhibits no distress when the mother leaves and avoids the mother when she returns. Mother and child appear "disconnected" and the child seems to prefer self-soothing over having the parent do so[9];
3. The ambivalently attached child is distressed when the mother leaves but is not soothed by her return. Caregiving of this type is described as "inconsistent," with periods of attention -- even "episodes of parental intrusiveness" -- followed by spans of time when the parent is inattentive or not available.[10] The child finds no safe haven in the parent[11]; and,
4. In the disorganized attachment category, the child looks afraid; she approaches the caregiver but then withdraws and may freeze

in place. The mother is frightening to the child, and the mother herself is often frightened.[12] The child feels both circuits simultaneously and does not know whether to move towards the mother for safety or flee, also for safety.

According to the research, about two-thirds of the children in the general population demonstrated a secure attachment; around 20% had an avoidant attachment; another 10 to 15% showed ambivalent attachment; and around 10% had a disorganized attachment.[13]

The attachment theory literature tells us that mild challenges help children learn and grow. It's only when those challenges are overwhelming or a parent repeatedly fails to restore a feeling of safety that attachment-related problems surface. Interestingly, a "mis-attunement" can be as simple as failing to anticipate what a child needs or underestimating a child's abilities.[14] Also, the attunement of a mother and child during the first year of life is a good predictor of how well a toddler will be able to control him/herself at age two.[15] Further, a parent's ability to attach reflects his/her own childhood and the parenting he or she received,[16] and attachment is a "powerful" predictor of how well we will parent our own children.[17]

Not surprisingly, the availability of this kind of consistent safety and security has serious and long-term consequences:

- Through attachment relationships, the parents' unconscious becomes the child's first reality.[18] In this way, early trauma, stress, and serious mis-attunement are handed down from generation to generation;
- Early attachment patterns affect *all* future relationships. They become the rules, scripts, heuristics, and schemas (models) we apply when we relate with others. They are unconscious and automatic, activate before we are consciously aware of them, and cloud, color, filter, and distort everything we see; and,

- Several large and long-term studies have charted the relationship between early caregiver love and attention, and later health and well-being. In the Harvard Mastery of Stress study, 91% of the participants who, 35 years earlier, reported they had not had a warm relationship with their mothers, were diagnosed with a serious disease (coronary artery disease, high blood pressure, alcoholism, etc.) in mid-life. One hundred percent (100%) of the participants who, 35 years earlier, rated *both* their parents as low in warmth and closeness were diagnosed with diseases in mid-life.[19]

There is another side of the attachment data that looks at the *parent's* experience of being parented; it's called the **Adult Attachment Interview (AAI)**.[20] Parents of secure attachment relationships have a "balanced view" and can provide a "coherent account" of their early childhood. Those in the dismissive or avoidant category struggle to recall their childhood and see themselves as "alone and on my own." Parents who had ambivalent or inconsistent attachment experiences find it hard to have a separate identity or achieve independence. And parents of disorganized attachment relationships are often conflicted, troubled, or even overwhelmed as a result of their experiences early in life. So it appears what's important is how well we *integrate* our attachment patterns.[21] In time, we will pair off and mix-and-match these various categories: early attachment experiences affect who we choose as partners, our capacity for intimacy, and the tools we bring to the relationship to cope with life's inevitable ups and downs.

ATTACHMENT AND SELF-CONTROL

Babies learn through experience whether when they are aroused, someone will be there to comfort them and make them feel safe. By providing mild challenges and being there to support the baby, parents teach **self-control**. The child, in effect, "borrows" the parent's prefrontal cortex"[22]: she models, through the attachment relationship, the parent's capacity for self-regulation. This begins with the parent's response to a baby's distress and then, though a slow, step-by-step process, is gradually generalized to other situations. During this time, it's the parent's job to shelter the child from becoming too aroused while at the same time, providing progressively more challenging experiences that increase the child's tolerance and confidence; all the while being there, providing a safe place for the child to return to and move from.

This step-by-step process applies to stress, fear, too much excitement, even over-exuberance: by comforting the infant or reining in a toddler, the parent "down-regulates" the amygdala and replaces arousal with safety. Self-control depends on the development of top-down networks that connect the **orbitofrontal cortex** -- and the **orbitomedial prefrontal cortex** in particular -- with the **amygdala**, as well as prefrontal connections to the **vagal nerve**, **hypothalamus**, and other areas of autonomic regulation.[23] As the cortex develops, it plays an increasing role in **inhibition**, beginning with a baby's involuntary motor movements: top-down networks from the cortex first inhibit many of the involuntary reflexes a baby is born with that are produced by the **brainstem**. This top-down circuitry then extends to include emotional impulses, inappropriate behavior, etc.[24] The child's success depends on how well the desired behavior is modeled, how many opportunities she has to fail and try again, and the extent to which she is able to hone her skills. As the top-down networks come online based on these experiences, the child gradually learns to exercise them herself. By internalizing the external regulation provided by early attachment figures, children find emotional balance which helps them become successful individuals as well as good social citizens.

ATTACHMENT AND SELF-ESTEEM

Children with secure attachments walk away with an important life message: you are loved, okay, and capable, no matter what. A securely attached child develops compassion, and learns self-acceptance, empathy, and moral reasoning skills. Children whose parents -- for whatever reason --cannot provide the nurturance they need to learn this message have to earn it, and can spend a lifetime doing so.

Parents love and attend to their baby unconditionally. Around the age of two, there's a shift from pure affection to affection and control, as a parents' job becomes teaching the child what is appropriate and safe. Shame is powerful tool for inhibiting behavior. Its power comes from its survival value, from the primal need to stay connected to the parent.[25] When a child does something wrong and is rebuked by her parent, disapproval makes the child feel anxious; since anxiety is unpleasant, the child learns to avoid undesirable behavior. The impact of shame is visceral, similar to being shunned or abandoned. When this feeling is quickly coupled with love and caring, the child learns to adjust her behavior. If, however, there's no rapid recovery back to "okay-ness," it can lead to hopelessness and despair. Shame is "**neurobiologically toxic**" for older infants, causing "permanently disregulated autonomic functioning" and a deep "sense of vulnerability" around others.[26]

Shame is not the same as guilt: guilt refers to a behavior; shame is directed at the person. The child comes to believe that somehow *she* is defective. An over-reliance on shaming in parenting undermines self-esteem: it can lead to a life of trying to prove one's worth or believing that only the threat of punishment will keep behavior in line.[27] Most people believe that because shame is so unpleasant, the very threat produces results. In fact, shaming does not appear to have preventive value: researchers have found that guilt is more likely to promote good, pro-social behavior whereas the more people are shamed, the "more likely they are to misbehave."[28]

ATTACHMENT AND RELATIONSHIPS

Through attachment, parents teach us how to engage with others... and the process may begin *biologically*. Steven Porges developed the **Polyvagal Theory** based on research grounded in the belief that "our autonomic nervous system unconsciously mediates social engagement, trust, and intimacy."[29] It does so through the "**smart vagus**."

Porges argues that in order for mammals to engage socially, we must first overcome our natural state of vigilance. To accomplish this, evolution has provided three adaptive strategies in the mammalian autonomic nervous system, each with its own distinct neural circuit[30]:

- The first strategy is **immobilization**. It depends on the oldest branch of the vagus nerve, the **vegetative vagus**. This unmyelinated portion of the nerve originates in the brainstem and, in the face of fear, the **parasympathetic nervous system** shuts the body down – we freeze;
- The second is **mobilization**, or fight or flight. When confronted with a fearful situation, the **sympathetic nervous system** increases metabolic activity and prepares us to flee or fight; and,
- The third strategy is **social engagement**. It depends on the "smart vagus," the **myelinated** branch of the vagus nerve, which calms the sympathetic nervous system, allowing us to stay engaged despite our natural desire to flee, fight, or freeze.

Porges' model links social engagement to attachment and social bonding through a series of stages. The theory says the nervous system automatically and unconsciously evaluates the environment to assess risk and based on that, chooses the appropriate strategy. If and only if there is a "**neuroception**" of safety – a *neural-based* assessment of safety – then social engagement can proceed. To do so, we must be able to immobilize *without* fear – to stay in one place long enough to engage, which depends on the smart vagus. The development of the smart vagus, in turn, depends on experience and in particular, the quality of our early attachment relationships.

The smart vagus is unique to mammals. The theory expands upon the previous two-circuit model where a child under threat simultaneously wants to flee and seek out the attachment figure for safety. Porges suggests that when we approach a social situation, we use the newest circuit first – the smart vagus – and if that doesn't provide the safety we need, "the older circuits are recruited sequentially."[31] According to the Polyvagal Theory, good enough parenting is said to increase our "**vagal**

tone," a measure of how well we respond to stress. Studies have found that people with higher vagal tone are able to respond more quickly, calm down more easily, and "tend to be more emotionally expressive."[32] People with low vagal tone are more defensive, impulsive, and irritable; they may exhibit behavioral problems as early as the age of three and often come from insecure attachment relationships.[33]

NO WORDS TO QUESTION, NO CAPACITY TO CHOOSE

Whatever we learn in our early relationships becomes our **attachment schema** – the unwritten rules, expectations, beliefs, and assumptions we carry in implicit memory about how relationships work. Implicit memories are "nonverbal, nonconscious, and embedded in the body…[they] cannot be recalled or explained away."[34] They are our learned history of what's dangerous and what's safe, the dominant pathways that become our mind patterns. And yet, the brain is malleable: insecure attachment can be repaired, just as secure attachment can be undone. Generally speaking, secure attachment has been found to be more resistant to change than insecure attachment.[35] Rats placed in an enriched environment, along with other interventions, have been shown to overcome the effects of inadequate maternal care.[36] On the other hand, when rats endure chronic stress or trauma later in life, the positive effects of early nurturance can be reversed.[37] Thus the **wisdom** of the brain is this: like attachment generally, **what happens to our brain is not always a matter of choice…or even under our control**. It's a lesson in compassion.

14

The Brain and Emotion

Emotions are more mysterious than action potentials, more speculative than models about how the brain works, and just as impactful as parenting and attachment schema. The limbic structures that support emotions drive "our hunger, lust, needs, and wants"; they also "move us towards others for safety, procreation, warmth, and connection."[1] We'll spend some time examining emotions because if we don't understand emotions, we can't understand people...and understanding people is what the helping professions are all about.

WHAT WE KNOW ABOUT EMOTIONS

We already know something about emotions:

- Processing in the brain is **distributive**: networks crisscross the brain and link functional areas to give rise to perception, cognition, and action;
- Emotion cannot be isolated from decision-making, intelligence, or behavior. The brain integrates emotions and memories into thoughts and action;
- Emotions are an example of the brain's embodied nature, immersed in a **brain-body-environment interaction**;
- Emotions can be preprogrammed -- as in the case of fight, flight, or freeze -- or learned; and,

- Much of our most important emotional lessons are learned before we are conscious of them, through our attachment relationships, including our ability to control our emotions.

Emotion can be defined as an *affective mental* response to a *stimulus*.[2] In some cases, the stimulus or situation affects us all in the same way: it has biological significance and our responses are automatic and reflexive.[3] Emotion can be said to "evoke motion," to prepares the body to act.[4] Emotion has also been called the foundation of reason: through emotions, we take in the external world, translate it into a physiological response, and then take the necessary action. Emotions register in the **autonomic nervous system**; they affect the inner body milieu and **viscera**. Changes in our physiology, in turn, alter behavior. In effect, we *feel* our way through life:

- Emotions have a clear evolutionary foundation: they build upon on an organism's most elementary pattern of movement – approach and avoid. In fact, "affective categorizations and responses are so critical that all enduring species have rudimentary reflexes for categorizing and approaching or withdrawing from certain classes of stimuli."[5] For creatures lacking a limbic system, approach or avoid are the only options available. Those with a basic limbic system and a **cingulate cortex** (the predecessor of the neocortex) can *feel fear*. And for mammals, which have a limbic system *and* a neocortex, far more complex emotions and greater range of responses are possible. Emotions guide, modify, and/or control our approach/avoidance instincts.
- Evolution has produced emotions because they have survival value. It has also preserved **emotional learning** so we can *remember* what we've learned. Early emotional memories are generally **implicit**, meaning they are unconscious or their source is unknown. Emotions allow us to communicate quickly with others: to warn

them of danger, alert them to an opportunity, and express our joy, sadness, or anger. It's thought that up to *90%* of an emotional conversation is non-verbal.[6] Since social situations have survival implications as well, emotions help us navigate social interactions where rule-based learning often doesn't apply.

- Emotions are markers of value, of what has meaning *to me*. We assign what's called **affective value**: the world begins as "an unlabeled place that we categorize by selecting what is of value and interest to us"; feelings, emotions, and emotional memories form "unconscious categories that are the source of *potential* meaning."[7] We feel our own beliefs, experiences, and opinions as *sensations* in the body. While not emotions per se, they have feelings associated with them which we experience in the **viscera**. When it comes to *other people's* beliefs, experiences, and opinions, we have to *construct* them: we have to ponder what they might be feeling, compare it to our own experience, and design what may be an appropriate response. We have to shift gears -- from an inner emotional experience to an outer theoretical one, which is hard work.
- Emotions are said to speak through the limbic system, although it's not a system per se. The orbitofrontal cortex is an essential part of emotional processing, but is technically not part of the limbic system. The hippocampus *is* part of the limbic system, but is not a key player in emotion. The insula and cingulate cortex are also highly interconnected with the limbic system, but are generally regarded as part of the cortex. Along with the amygdala, these structures depend heavily on the **hypothalamus**, which is not part of the limbic system and yet controls the physiology of emotions. As a result, it's more accurate to think in terms of the many regions of the brain that participate in emotion and emotional processing.
- Emotions are **patterns of arousal**. They are influenced by needs, subcortical preferences, representational memory, expectations, beliefs, assumptions, etc. We already know the prefrontal cortex

interacts with the amygdala: research has shown that "at low levels of stimulation," emotions play a mostly "advisory role"; at medium levels, emotion and self-control compete in a battle for short-term supremacy; and, "at high levels of stimulation," such as when we are physically or mentally exhausted, cortical processing can be overrun by emotion.[8] Once a pattern of arousal is triggered, it's very hard to restrain. The **wisdom** of the brain thus offers us this insight: while humans *can* shift from reaction to willful control, **choosing takes effort and there's no guarantee we'll make a good choice.**

MORE ABOUT EMOTIONS

Emotions generally happen in response to something: the stimulus can happen in the environment, like the sight of a scary dog, or be internal, as in a happy memory. **Basic emotions** include happiness, anger, fear, disgust, sadness and surprise. They share what's called a "**stereotyped expression**" across species – e.g., bears and people look the same when they're angry. There are also "**social emotions**" which include embarrassment, guilt, shame, jealousy, hate, and contempt. Some emotions are considered hardwired or instinctive. Nevertheless, while we share the same general kinds of emotions, "each of us still has to develop his or her own individual classification" of what's good and bad, and in this way "we develop a personal version of the world."[9]

Emotions are often distinguished from feelings: **emotions** are the autonomic physiological responses that happen in the body, while **feelings** involve the conscious awareness of those bodily responses – how the brain accounts for, explains, or describes them. Emotions are said to perform three basic survival functions. First, they notify us of something important: it can be physical, psychological, or social. Second, they focus our attention, motivating us to deal with the situation in some way. And third, they produce changes in the body to prepare us to do what needs to be done and maintain our attention until the event or stimulus is resolved or passes.[10] The experience of a given emotion is generally unlearned, automatic, and recognizable from person-to-person. *Within* a person though, emotions vary by intensity and duration; emotional triggers and thresholds, as well as the emotional load we can bear, vary from person-to-person as well.

Emotional processing depends on the complex circuitry of the **orbitofrontal cortex, amygdala, sensory systems, anterior cingulate cortex, insula**, and other cortical and subcortical structures. An emotion "begins" with an **appraisal**, an automatic and generally unconscious evaluation of an **emotionally competent stimulus**. Our appraisal is not the stimulus itself, but our *interpretation* of it. The stimulus can be present or called up from memory; its competence can be innate or learned, a threat or an opportunity. Representations of the stimulus are processed through one or more of the brain's sensory systems, and information is also passed off to various emotion-generating sites in the brain, such as the amygdala and the orbitofrontal cortex.[11] Once the brain detects the stimulus, it transmits command signals to the endocrine glands to secrete hormones that affect the body tissue and the brain; to the **autonomic motor system** to handle physiological changes in the cardiovascular system, visceral organs, and inner body tissue; and to the musculoskeletal system to address such outward features as fight, flight, or freeze, bodily posture, and facial expressions.[12] These signals translate the appraisal into physiological changes in the body, into an **emotional response**. To create the *feelings* that go along with an emotional response, signals from these body states are then sent back to the brain where they are mapped in the cortex and the **interoceptive** (the perception of the body's interior) signals are brought into conscious awareness.[13]

THE MANY PLAYERS IN EMOTION

Many parts of the brain participate in emotion; no single area is responsible for any one emotion or its processing. The **amygdala** is a critical player in emotional processing. It works in conjunction with the orbitofrontal cortex to produce the basic emotions. The amygdala is also responsible for encoding the emotionally salient features of an event or stimulus; storing those features for future reference; and producing the unconscious autonomic responses triggered by those features when faced with the same or a similar situation in the future.[14] While the amygdala responds to a variety of emotions, it is especially attuned to fear. In addition to the amygdala, many regions of the brain participate in emotion:

Orbitofrontal Cortex and Orbitomedial Prefrontal Cortex

- Enables affect regulation and helps organize behavior
- Translates complex social information such as gestures, eye contact, etc., into meaningful information
- Coordinates and helps balance the sympathetic and parasympathetic nervous systems
- Serves as a storage site for circumstantial information about an emotional event[15]
- Monitors whether emotional decisions are appropriate

Cingulate Cortex

- Helps sustain engaged attention when dealing with a difficult problem
- Plays an important role in emotional self-control
- Responds to changing situations and recognizing when current behavior is not meeting desired ends
- Participates in maternal instincts and emotional stability
- Anterior cingulate cortex plays a role in mentalizing about others
- Anterior cingulate cortex works with insula in feeling physical pain and social pain (hurt)

Insula

- Helps link bodily sensations with emotions
- Highly sensitive to stimuli that trigger a strong visceral reaction

- Plays a primary role in fear, anger, disgust, happiness, and sadness
- Works with anterior cingulate cortex when we or a loved one experiences pain

Somatosensory Cortex

- Participates in "gut feeling" – activates associated implicit memories and helps guide decision-making
- Provides information about bodily experiences
- Plays a role in social feelings

Hippocampus

- Participates in the organization of emotional learning and memories
- Associates context and factual information with emotional memories

Septal Area

- Works to inhibit aggression, decrease fear
- Thought to help mobilize empathy into action

Cerebellum

- Adjusts laughter and crying to the situation

Hypothalamus

- Controls the physiology of emotions: translates experience into bodily states

- Plays a key role in the hypothalamic-pituitary-adrenal axis (HPA) and the autonomic nervous system
- Involved in regulating aggression

Right Ventral Lateral Prefrontal Cortex (rvLPFC)

- Involved in taking another person's point of view
- May be involved in the inhibition of beliefs

Ventral Medial Prefrontal Cortex (vMPFC)

- Participates in social emotions such as pride and guilt
- Activated prior to a decision, by tasks involving rewards and punishment
- During an emotional response, helps regulate attention, influences which representations are retrieved from memory, and shapes mental plans to respond to the situation

Right Ventral Prefrontal Cortex (rvPFC)

- Participates in physical pain, distress, and regulating social pain

Ventral Striatum

- Involved in the prediction of rewards

Periaqueductal Gray (PAG)

- Plays a role in pain, attachment behavior
- Involved in laughter and crying, freezing and running in response to fear, as well as expressions of disgust and fear

Bed Nucleus of the Stria Terminalis (BNST)

- Activates lower-level arousal in the form of anxiety or unease

The many regions of the brain involved in emotion support many different types of emotions. For example:

- *Jealousy*. Jealousy is thought to be the combination of fear, anger, betrayal, and sadness;
- *Anxiety*. Anxiety is defined as when working memory is flooded with fretful and worrisome thoughts. It differs from fear in that anxiety is fear of what *might* happen. Anxiety acts as an *avoidance* motivational system. Amygdala dysfunction is thought to underlie all anxiety disorders.
- *Anger*. Anger involves the right orbitofrontal cortex, anterior cingulate cortex, amygdala, hypothalamus, and periaqueductal gray. While its expression is assumed to be cathartic, doing so actually increases feelings of anger rather than dissipates them. Anger is automatic and related to frustration or feeling trapped. It acts as an *approach* motivational system. The amygdala plays a key role in both anger and aggression, and the amygdala and superior temporal sulcus (STS) are involved in the detection and processing of implicit anger. Anger may be represented by greater activity in the left hemisphere.
- *Sadness*. Sadness is a *withdrawal*-related behavior. It is supported by the amygdala and right temporal pole.
- *Disgust*. Disgust is related to smell and the olfactory system. It is supported by the anterior insula and anterior cingulate cortex in avoidance behavior, and the insula and basal ganglia to recognize disgust. Disgust may be unique to humans.
- *Aggression*. Aggression is related to fear. It involves the amygdala, hypothalamus, and periaqueductal gray. It can be activated to defend a person, one's family, possessions, or belief system. It can also occur in response to disrespect or an affront to one's self-image.

Fear is special; it is a dominate force in human nature. We have an **early learning bias** towards fear and it may be a baby's first emotion. Parenting plays a pivotal role in shaping the safety-danger continuum and as a result, we each have very different thresholds for emotions like fear, anger, and joy, and these thresholds are heavily influenced by both genetics and early learning.   As opposed to anger, fear and defensive aggression lead to withdrawal behavior, while anger leads more to approach behavior. Fear can cause isolation, negativity, and all forms of reactivity. It lies at the heart of much human behavior. Even happiness…may be the absence of fear.

15

The Many Faces of Emotion

As helping professionals, we bear witness to a great deal of emotion. In the neuroscience of emotion, some of what we learn reinforces what we know and validates our approach. Other things are more mind-shifting and require some time to assimilate. Whenever we look into the emotional state of a client, student, etc., we should see the brain doing its job: emotion helps the brain make sense of the world, maintain a coherent sense of self, and sustain some semblance of control...especially in front of others.

EMOTIONAL OPERATING SYSTEMS

One way to think about emotion is using Dr. Panksepp's primary emotional operating systems -- the Seeking Systems, Rage System, Fear System, and Panic System. We share these basic systems with all mammals; each has survival value and all are open to new learning which makes them especially useful[1]:

- The Seeking Systems are related to experiences such as anticipation, appetite, desire, or want, and the emotions of interest, feeling energized and excited, curious, and stimulated. One subsystem of the Seeking Systems, the lust circuitry, highlights emotions like love, delight, satisfaction, joy, and contentment;
- Anger is derivative of the Rage System. The Rage System is triggered by anything that restricts freedom like being caged, feeling

trapped, lacking options or the opportunity to choose, etc. Although there are predatory and inter-male circuits in addition to the Rage System, only the latter involves the experience of anger[2];
- The Fear System produces emotions ranging from mild to chronic anxiety, distress, anguish, fright, terror, and a state of dread; in the body, it manifests as worry, nervousness, tension, and worse.[3] It also controls pain sensitivity especially during a fearful event, enabling us to feel the pain of an injury only after the event is over[4]; and,
- The Panic System is involved in both physical and psychic pain, and generates feelings of panic-based anxiety, separation anxiety, loneliness, grief, sorrow, and loss.

Fear is a foundational human emotion and as a result, it receives a lot of attention. The terms "fear," "threat," and "danger" are often used interchangeably. Fear is highly **generalizable**: we can learn to link any thought or event with fear, or any thought or event that *vaguely resembles* the original one. Survival is also served by the fact that fear is easily learned. Fear can be learned through direct experience: in **conditioned fear**, when a neutral stimulus such as a buzzer is paired with an aversive event like an electric shock, the subject quickly learns to react to the buzzer *as though* it had been shocked.[5]

Fear can also be learned through instruction or by observing a fearful event. If you've ever been bit by a dog, you probably learned to be afraid of dogs. If you told a friend about how terrifying that experience was, he too may have learned to be afraid of dogs simply based on what he was told. Similarly, if a sibling sees you getting a shot at the doctor's office, you don't need to tell him it hurts – he already knows. And the experience doesn't have to be physical – we can learn to fear other cultures and people. Fear conditioning is an example of **implicit emotional learning** because the learning is "expressed indirectly" through one's behavior or physiology.[6] Fear that is learned through instruction or observation is an example of **explicit emotional learning**. Explicit emotional learning depends more on the hippocampus, implicit depends more on the amygdala.

Because it is so closely linked to survival, emotional data is given high priority in the brain. However to be useful, emotional events must be remembered. The **amygdala** encodes emotional events -- the implicit parts that have "high emotional content" – and serves as "a memory system" for such information.[7] When evaluating a new stimulus, the amygdala compares it against records of past events. If it's the same as or resembles a past stimulus, the amygdala triggers a series of unconscious changes in the body to deal with the situation. The amygdala's appraisal is both unconscious and automatic; it is another example of how the past keeps shaping our experience of the present. The **hippocampus** creates the memory of the explicit or known facts *surrounding* the emotional event. The information is stored in the neocortex, in the **orbitofrontal cortex**: the orbitofrontal cortex stores the circumstances, the amygdala stores the fear.[8]

EMOTION AND THE CORTEX

As we've seen, many parts of the brain play a role in emotion…including the cortex. The cortex works with other parts of the brain to focus attention, overrule emotional reactions, navigate social situations, and devise ways of handling difficult problems. Yet the cortex is no panacea -- it's just as likely to unnecessarily distort or amplify an emotional state. Consider the following examples:

- Thanks to the cortex, human beings can use symbols to make sense of the world. We also use symbols to react to less than real emotional events -- to hand gestures, for example. Similarly, through **working memory** (supported by the dorsal lateral prefrontal cortex), we're able to keep replaying past events and feel their emotional pull. Humans have an amazing capacity to recall *and suffer* over the past, sometimes decades after the fact.

- In a "*cognitive* appraisal," we can talk ourselves into a particular emotion, such as a panic attack or anxiety, or sabotage a well-intended effort. Beliefs too, show how thoughts trigger emotions – consider a heated discussion about religious convictions, abortion, immigration, wealth distribution, etc. Also, recall that the "drive to defend" is triggered by a perceived threat to a person's body, his/her possessions, bonded relationships, or to one's "**cognitive presentations**."[9] The fact is we can get pretty worked up over thoughts, especially when they're part of who we think we are – part of our sense of self.
- The cortical ability to project into the future and imagine future events allows us to create an entire array of "what if" scenarios. Most of these will never materialize and many haven't a chance of materializing; we call this habitual tendency "worry." In addition, too much intervention or involvement by the prefrontal cortex can seriously dampen **creativity** by overthinking a given situation or imposing too many rules.
- One of the "most striking features of emotion is the profound variability among individuals in the quality and intensity of their response to *identical* stimuli" (emphasis added).[10] Repetitive, irritating but minor stressors can pile up in the cortex and manifest in the body. It's said that our body reflects our emotional history -- that it speaks our mind: shoulder pain can come from carrying too many burdens, tennis elbow from a persistent mental or emotional irritation, back pain from survival and security worries, etc.[11]
- When we don't regulate our emotions well, the brain resorts to coping strategies to try to maintain a semblance of control. **Defenses** are such a strategy -- they distort reality. While the distortion itself takes place through implicit memory, the cortex provides the "rationalizations and beliefs" that support the defenses and uses them to organize our experience, sometimes for a lifetime.[12] The cortex also provides us with our own ongoing mental dialogues:

when these dialogues misrepresent reality, we can keep trying to assign blame, shut out alternative explanations, cut off possible solutions, or look outside for causes.
- Questions remain about the role of **consciousness** in emotion: whether an emotion or feeling occurs through consciousness or only unconscious association is involved. Similarly, most appraisal theories claim that emotions *start* with an appraisal; that the content of the stimulus is appraised, it is associated with memories, and then, as a consequence, changes take place in the body and in behavior.[13] It's likely, however, that emotions and feelings travel *both* conscious and unconscious routes, and appraisals and emotions "blur" into one another. Regardless, one never causes the other...the relationship is at best, a correlation.

In sum, we can't count on the cortex to come to the emotional rescue. Sometimes, it gets overruled; other times, thinking makes matters worse. While cognitive and emotional processes interact in the brain, "the flow... remains balanced only in nonstressful circumstances. In emotional turmoil, the upward influences of subcortical emotional circuits...are stronger than the top-down controls."[14]

EMOTION AND THE AMYGDALA

The amygdala serves as the hub of our **threat and danger detection system**. It is comprised of thirteen interconnected nuclei, the most well-known being the lateral, basal, and central nuclei. In terms of function, the **lateral nucleus** interfaces with the senses; the **central nucleus** controls the expression of emotional reactions and their associated physiological responses, and also forms fear associations[15]; and the basal nucleus works with the striatum to control action-related behaviors.

While the amygdala responds to many different types of emotion, its primary role is processing fear and detecting threat.[16] Not surprisingly, **projections** to and from the amygdala are so numerous that it has the potential to influence the nervous system "from the spinal cord, brainstem and hypothalamus to cortical regions in the frontal, cingulate, insular, temporal, and occipital cortices."[17] The amygdala is reciprocally connected to all the **sense modalities** through the thalamus, and directly with the **olfactory system;** also, to the **brainstem** and the **hypothalamus** (for autonomic, visceral, and defensive functions, among others), the **basal forebrain** (which is related to aggression and motivation), the **hippocampal formation** (which is involved in learning and memory), and while not reciprocal in nature, to the **bed nucleus of the stria terminali** (which is related to anticipated and generalized anxiety), and the **striatum** (which assists in integrated motor responses, as well as reward and goal-directed activity).[18]

With respect to the neocortex,[19] projections *to* the amygdala come mainly *from* the orbitofrontal and the medial prefrontal cortices which work with the amygdala to assign emotional value. The amygdala also projects *to* the orbitofrontal cortex, the mediolateral cortices, and the **premotor cortex**, among others. The **insula** provides one of the "strongest cortical inputs" to the amygdala. The amygdala also sends and receives projections to and from the **cingulate cortex** and **multimodal** areas of the **temporal lobe;** projects "lightly" to the **parietal lobe**; and receives most of its visual input from the ventral "what" path of the visual system.[20] These connections position the amygdala at the hub of the danger detection system, capable of launching a bodily response in the face of a threat.

In addition to fear, the amygdala is now known to be involved in processing facial expressions and making social judgments about those faces; processing rewards and their use as a motivational or reinforcement tool[21]; categorizing – that is, finding **patterns** among people and placing them into groups[22]; and recognizing and responding to "socially salient stimuli."[23] It's also been found to respond to positive visual scenes as well as joyful voices,[24] and is involved in "gaze monitoring, food selecting, sexuality, and much more."[25]

EMOTION AND DECISION-MAKING

Emotion makes decisions possible. Emotion allows us to discern what's good or bad, what has value and meaning, and to do so *prior* to deciding.

Dr. Antonio Damasio's **Somatic Marker Hypothesis** describes the process through which emotions create optimal decisions. It goes something like this[26]:

1. When faced with a decision, we often react to the situation emotionally;
2. The physiological changes associated with this reaction are "somatic markers";
3. The orbitofrontal cortex is thought to help link somatic changes with past experience, and, in conjunction with other parts of the brain, sort through previous situations that triggered similar patterns of arousal;
4. Once the orbitofrontal cortex identifies several matches, it uses these situations to evaluate the choices currently available to assess which might lead to a positive result; and,
5. Decision-making can then proceed by *selectively* attending to those options that have the greatest likelihood of a positive outcome.

It was largely because of Damasio's work that emotion gained in stature in the neuroscience community. More recent research has modified his work suggesting that rather than somatic markers, the orbitofrontal cortex likely applies **social knowledge** to our decision-making process; recognizing when, for example, it might be appropriate to give a client a hug, cry with a student, or outwardly express frustration.[27]

Understanding the role emotion plays in decision-making is important for helping professionals because so many of our professional interactions involve emotions-based decision-making. What, for example, constitutes a win-win? It's the feelings and perceptions of the people involved. Because it depends on *what's important to them*, it may reflect the process, the agreement, the relationship, or something else. One person may feel it was a win-win because she got something she wanted; another because he was able to save face; a third because everyone walked away able to work together again; and a fourth because he got something of value to the *other* people and he wanted to be seen as a winner. Regardless, understanding the role emotion plays in decisions is critical to understanding people and what makes them tick.

Recommended Reading
Affective Neuroscience, by Dr. Jaak Panskseep

A
Closing Thought

A Neuroscience Question: How real is reality?

We've said the brain creates a picture of reality we live in and through, often without question. We also know the brain pays attention to some things and not others, and takes in only a sliver of what's going on. We've learned what the senses convey is not a direct transmission, but the product of a complicated processing system. Moreover, the visual processing system, which is a key part of perception, *constructs* what we see – it applies a set of known rules about how the world works and then literally "guesses" based on past experience. So how real *is* reality? Consider the following:

- The **corpus callosum,** which ties the two hemispheres of the brain together, actually unifies perception by melding together the two fields of vision (one from each eye), only a portion of which overlap;
- Sensation is considered an abstraction because the brain can only interpret what it takes in within the confines of past experience;
- Through **reentrant pathways**, higher processing centers in the brain reformulate incoming sensory information by feeding it back to lower processing levels[28];
- The brain must first deconstruct an experience to process it. When the brain reassembles it, it adds in emotions, drives, instincts, as well as the rules, biases, and expectations we've formed over a lifetime; and,
- The brain automatically includes our wants, desires, and dreams in our perceptions: "Our internal computations, which we believe to be objective…are implicitly colored by who we are and what we are after."[29]

There was a time when science assumed we perceived the world as it exists. This is known as naïve realism. But the **wisdom** of the brain is that **the brain and the mind exert a powerful influence over our sense of reality**. In truth, there is no such thing as color or tone, smells or tastes *outside the brain* -- there are electromagnetic waves, pressure waves, and chemical compounds out of which the brain constructs sights, smells, etc.[30] So how real is reality? Or should we say: How real is *your* reality? It seems quite possible that what we experience as reality is a *construction*; that the brain fills in the gaps and makes everything appear real, allowing us to feel much more certain and in control than we actually are.

16

Emotional Reactions

In an emotional situation, it's often said we can either react or respond. We often **re-act**: we fall back on an established set of patterned, automatic, and reflexive actions that we repeat. This is especially true in the face of fear. Yet, because of how the brain processes information, the choice to re-act or respond is not really an either-or situation...and, as is often the case, our ultimate behavior depends on a whole host of factors.

CONDITIONED FEAR

Parents of every species teach their offspring about fear. For human babies, parents shape the **safety-danger continuum** – they help form the child's emotional set-points and build the neural architecture for storing emotional memories. By design, the brain is set-up to process fear quickly and generalize from the experience...and it takes effort to inhibit fear. In effect, our fears have served us well: "One reason that our instinctive fears have not been updated is that...we can learn any new fear we need,"[1] thanks to the cortex.

Much of what we know about fear is due to neuroscientist Joseph LeDoux's work with **conditioned fear**. LeDoux describes the fear system as a defensive system that maximizes our chances of survival.[2] It is programmed by evolution and thought to have developed *independent* of conscious awareness,[3] meaning at least from the standpoint of evolution, *being* afraid is more important than *feeling* afraid. At the hub of LeDoux's model is the **amygdala** which specializes in identifying personally relevant emotional information and using past experience to appraise the meaning of a stimulus and whether it's a threat. If it is, the amygdala signals the **hypothalamus** to launch a bodily response.

LeDoux was interested in how stimulus-related information got to the amygdala in the first place. He found that it travels two separate paths:

- On the **low road** or fast circuit, incoming sensory information travels to the **thalamus** which sends it directly to the amygdala. This information is basic and crude, but the path is fast and direct.[4] Generally speaking, the input is sufficient for the amygdala to assess whether the stimulus roughly resembles previous stimuli.
- Incoming sensory information from the thalamus is also sent to the primary cortex for that particular sense modality, for further analysis. The primary sensory cortex then conveys its results to the amygdala. This is called the **high road** or slow circuit. Contextual information about the stimulus also travels the high road. This path takes longer but provides more refined and meaningful information.[5]

According to LeDoux, information about the stimulus travels these two paths *simultaneously* and *both* converge in the amygdala. Using **implicit** and **explicit memory** sources, the low road shapes behavior based on past experience and the high road compares incoming information to similar events and passes on its analysis to the amygdala. It's said the low road "primes" while the high road "confirms." The high road is *not*, however, a "conscious road to the amygdala"[6]; all of this takes place outside of, and prior to, conscious awareness. It takes the amygdala less than 50 milliseconds to react to a potential threat, while it takes between 500-600 milliseconds for an experience to enter conscious awareness.[7] By the time a threatening situation reaches conscious awareness – *if* it reaches conscious awareness -- the brain has already decided what to do…and done so based on early learning, implicit memories, and past experiences, *not* conscious deliberation.

It's helpful to think of the process of evaluating a stimulus as an *iterative* one: while the amygdala re-acts, the cortex continues to feed information to the amygdala to refine and adjust its response. Some researchers have suggested that activity in the amygdala might reflect both "initial associatively-driven **automatic** evaluations" and "subsequent **reflective** evaluative processing."[8] As a result, rather than replace an initial evaluation, the evaluation is "continually updated and integrated with additional attitudinal, situational, and motivational information to generate increasingly complex evaluations."[9] For the amygdala, what's at issue is whether to launch a bodily response to a threat. The upside of the low road is that it ensures we err on the side of caution. The downside is that it can cause mistakes: crude, re-active information can lead to stereotypes, rigidity, and broad-brush categorizations with little real experience to back them up. To paraphrase LeDoux: if, "for genetic or experiential reasons," the lower-order pathways are better at triggering the amygdala than the higher ones, "we would expect those individuals to have rather limited insight in to the nature of their emotional reactions."[10]

The ability to inhibit fear (or any other emotion) and assert self-control depends on the strength of the top-down, cortical-limbic circuitry (See Chapters 12 and 13). This circuitry is **experience-dependent** – its strength depends on the opportunities we've had, especially early in life, to successfully practice self-control and inhibition. It's important to understand that the connections *from* the amygdala *to* the cortex are far stronger than those from the cortex to the amygdala.[11] There also appears to be a pattern of "**reciprocal activation**" such that as activation in the cortex increases, activation in the amygdala decreases, *and vice versa*.[12] Moreover, some important areas of the prefrontal cortex (PFC) don't project to the amygdala *at all*, including parts of the **dorsolateral prefrontal cortex** which focuses more on **executive functions**.[13]

All of this means that the amygdala is in a better position to directly influence higher cognitive functions than vice versa. It also means that executive functions related to goals, planning, conscious attention, etc. can only influence the amygdala *indirectly* since no direct projections from the dorsolateral prefrontal cortex exist. This starts to explain why we are so easily consumed by our emotions and it's so hard to regain control once that happens. Once a stimulus is associated with pain, threat, danger, or fear, the linkages are almost impossible to **unlearn**. Unlearning fear requires the same **synaptic plasticity** that learning does. But rather than extinguishing a fear, new learning preserves the memory while inhibiting the arousal networks that mobilize a threat response.[14] The net result is that we never really forget, and the original memory can easily be reactivated under stressful conditions.

In terms of conditioned fear and its implications, LeDoux cautions us: "Fear conditioning may not tell us everything…about fear, especially human fear…The neural circuits involved in responding to conditioned fear stimuli may participate in, but are probably not sufficient to account for, more complex aspects of fear-related behavior, especially [those involving] abstract concepts and thoughts…"[15] He states that while it's tempting to see the amygdala as the "centerpiece of an emotional system of the brain…We know far too little about the neural system-mediating emotions other than fear and far too little about variants of fear other than simple forms of conditioned fear."[16]

THE THREAT RESPONSE

We are now in a better position to understand how the **threat and danger detection system** works:

1. Fear generally happens in response to something – it can be physically present or recalled from memory, an imminent threat or a perceived one;
2. Incoming sensory information about the stimulus travels the low road and the high road to the amygdala. The low road provides "quick and dirty" information to determine whether it roughly resembles a prior stimulus. The high road feeds more refined information, including contextual information;

3. The amygdala conducts an initial appraisal and, if the threat is the same as or resembles a past stimulus, the activation of the amygdala triggers the "automatic activation" of arousal networks.[17] This mobilizes a **threat response**, which often "takes the form of an escape"[18];
4. The amygdala activates networks in the **brainstem** and the **hypothalamus** to provide energy for the body to respond and translate the threat appraisal into physiological changes, respectively. The hypothalamus is the controller of the **autonomic nervous system** and is also at the helm of the hypothalamic-pituitary-adrenal axis;
5. Through the autonomic nervous system, the **sympathetic nervous system** triggers fight or flight. Fight or flight can be thought of as "shorthand" for any activity which "requires high metabolic voluntary muscular activity."[19] The sympathetic nervous system also diverts energy away from activities like digestion, fighting off a virus, and thinking deep thoughts, and towards preparing the body to act;
6. Projections from the **central nucleus** of the amygdala (See Chapter 15) cause the hypothalamus to activate the pituitary gland. The hormone released by the **pituitary gland** causes the **adrenal glands** to release glucocorticoids, including **cortisol** and other stress hormones. **Adrenaline** is also released to ready the body for action;
7. Taken together, these physiological changes translate the threat into the body's response in the form of fight or flight, increased blood pressure, heart rate, sweating, etc., and the release of hormones into the body. Meanwhile, the **cortex** continues to send messages to the amygdala, informing it as to the precise nature of the threat[20];
8. More often than not, the threat is temporary; it passes, is responded to, and is resolved. Once the threat has passed, amygdala activation decreases and the **parasympathetic nervous system** takes over; it slows our breathing and heart rate, relaxes our

muscles, and brings the body and mind back to what constitutes a normal state; and,
9. *All of this is completely normal.* The emotionally salient aspects of the event are filed away by the amygdala, and the facts are processed into long-term memory by the hippocampus and stored in the orbitofrontal cortex. Overall, the threat and danger detection system performed as designed – it maximized our chances of survival.

The threat response is, in many cases, **a threat re-action**, the difference being that a re-action happens *reflexively*…it is not under our conscious control. Whether or not a threat response is launched depends on a multiplicity of factors. For example, as we've said, threat, safety, and stress are *personal*: each of us has our own threshold for the emotional load we can bear. In addition, not all stimuli are alike: we can feel a little afraid, or threatened for an instant; the hypothalamus can be partially aroused. There are also multiple networks at work: since the **high road** keeps feeding more refined sensory information about the stimulus to the amygdala, that information can, in time, interrupt a threat response. Similarly, while the amygdala is busy launching a threat response, **inhibitory forces** are at work. The connections between the amygdala and the orbitofrontal cortex (OFC) and orbitomedial prefrontal cortex (OMPFC) are "dense bidirectional networks." As emotional and visceral information flows into the cortex, the cortex works to regulate and adjust the output of the amygdala on the hypothalamus,[21] thereby mitigating the response. There is also the moderating input of the **smart vagus**, another aspect of our emotional regulation that depends upon early learning and experience (See Chapter 13).

Finally, in launching a threat response, the autonomic nervous system may have more options than simply fight or flight. Most of the research on fight or flight has focused on males and male rats in particular.[22] While fight or flight may be adaptive for males, females have different concerns: they need to protect their offspring.[23] Accordingly, animal research has found that females may **tend and befriend** more than fight or flight. Researchers suggest that while males lean more towards *physical* aggression, females show "as much or more" *indirect* aggression in the form of gossiping and rumor-spreading.[24] They point to human studies that show, under threat, "the desire to affiliate with others is substantially more marked among females"...and this represents "one of the most robust gender differences in adult human behavior."[25]

STRESS AS RE-ACTION

Once the threat passes, the body and mind return to a more normal state. When there is no natural flow back to a calm, relaxed state, this otherwise normal cycle then becomes **stress**, which is particularly harmful when it's chronic. With chronic stress, all systems are on high alert and the sympathetic nervous system is caught in a recycling loop. The body is flooded with stress hormones like cortisol, adrenaline, and norepinephrine, and keeps our **subjective inner milieu** in constant turmoil. Over time, chronic stress can cause the body to deplete nutrients more rapidly and weaken our immune system.[26]

Like a threat response, the likelihood and level of stress one experiences depends on a variety of factors. In addition to having our own set-point for stress, stressors differ in intensity. There are the everyday **transient stressors** we all face: uncertainty, problems at work, financial concerns, family squabbles, etc. Stress of this nature depletes energy and can make creativity and new learning difficult. **Chronic stress** is prolonged, as the sympathetic nervous system is continually over-activated and we become stuck in a heightened state of vigilance. **Traumatic stress** is even more extreme, where some past event keeps infecting our present-day experience. In this case, massive amounts of glucocorticoids are circulating in the body, and "long after a traumatic experience is over, it may be reactivated at the slightest hint of danger...[precipitating] unpleasant emotions, intense physical sensations, and impulsive and aggressive actions."[27]

Stressors also vary in duration. Short-term stress can be stimulating and is essential to learning. Hopefully, everyone has had the experience of finally figuring out an algebra problem or getting the Zumba steps right…it feels good! Long-term or chronic stress decreases learning, and sustained high levels of glucocorticoids are known to cause neurons in the hippocampus to atrophy.[28] In the case of traumatic stress, the amygdala of trauma victims grows hypersensitive and hyper-reactive because the sustained level of glucocorticoids in their system has caused the synapses in the amygdala to undergo **long-term potentiation** (See Chapter 5). Trauma victims live on edge: "The core features of traumatic stress [are] timeless reliving; re-experiencing images, sounds, and emotions"… it is "expressed not only in fight or flight but also as shutting down and failing to engage in the present."[29] Massive amounts of glucocorticoids also kill neurons.

AN EMOTIONAL ANALYSIS

To summarize the topic of the brain and emotion, let's consider an example. You're at a conference, attending a session on the neuroscience of emotion, and the presenter states the following:

> "There's a deadline looming, some divisive issue, or some other tense situation and instead of collaborating, people are working at cross purposes. Instead of listening, they're arguing. Instead of thinking, they're defensive and posturing. Why is this? Well, from a neuroscience perspective, we have three brains molded into one, a triune brain. When we're feeling under threat, the limbic system or mammalian brain sends a signal to the body that we're not safe, we're in danger. The brainstem, or reptilian brain, triggers a fight or flight response while the central nervous system floods the body with stress hormones. The third part of the triune brain, the prefrontal cortex, allows us to think and reason through a situation. Unfortunately, the limbic system's call for alarm overrides the prefrontal cortex and with the thinking part offline, our reptilian brain takes over: the higher brain shuts down and we lose our ability to listen and think, and take in new information.

We react based on our primitive survival needs. It's a predictable hard-wired response to threat. It's inevitable."

What's your analysis? It could be argued that this presentation misses the subtlety of the **body-brain-environment interaction** and doesn't give the audience the whole story. Now you understand that:

- *Not all stimuli are alike*: some may be novel, others specific, while some are more complex and abstract;
- *Not all stimuli generate a threat response*: they may trigger anger, suppression, or avoidance, and what we know about fear may or may not apply to these circumstances;
- *Not all fear is "conditioned fear"*: the neural circuitry of conditioned fear may not pertain to more complex forms of fear, especially in humans;
- *Not all threat responses are the same*: the hypothalamus can be partially aroused, and responses can be modulated or adjusted downward, interrupted, or overruled;
- *The prefrontal cortex (PFC) isn't 'offline'*: under conditions of fight or flight, the PFC does not get all the energy it needs because it's being diverted to more crucial matters;
- *Fight or flight aren't the only options*: submit, tend and befriend might be possible as well, especially in females; and,
- *When the orbitofrontal cortex (OFC) is overwhelmed, chances are it is temporary*: the orbitofrontal cortex can re-assert itself during a threat response.

Finally, *a threat response and stress are not synonymous*: while some texts use the terms interchangeably, in a threat response, the sympathetic nervous system and related hormonal releases accelerate the body and prepare it to escape. When the fear passes, the parasympathetic nervous system restores the body and mind to a state of personal balance. Stress typically involves prolonged activation of the sympathetic nervous system.

Chronic and traumatic stress can cause the body and the brain to begin to break down.

And therein rests the **wisdom** of the brain: ***it's important not to try to control the expression of emotion***. Emotions are markers of value; they point to what people care about, what matters most to them. They are one of the ways humans communicate with one another. To assert control over or in some way mute the expression of emotion undermines the purpose of having a conversation. Even at a low roar, people are still communicating; up to 90% of an emotional conversation is thought to be non-verbal.[30] The point is not to be consumed by emotions or spend an entire session simply venting.

Recommended Reading
The Body Keeps the Score, by Dr. Van der Kolk

17

This Too Shall Pass

Parenting, emotion, and life experience help shape the neural connections that give us our individual nature. Parents, in particular, are our first and primary teachers. Hopefully every child has been taught how to learn and better still, to love learning. Consider the relationship between learning and early life lessons:

- Learning requires safety: people need to feel "physically safe and emotionally secure" before they can pay attention to new material[1];
- The more we learn, the more we're able to learn: because of our tendency to find patterns and draw associations, the more associations we have, the more "hooks" we have to attach to new learning[2]; and,
- Learning is directly related to emotions – how one feels about a "learning situation" can either encourage or inhibit learning.[3]

Parents teach us labels that structure reality and are filed away in memory – in patterns of connections dispersed through the brain. According to the **computational theory of mind**, we use memory to assemble perceptions into **representations** which become the basis for learning *new* representations and forming *more* associations. They also provide the framework for preserving new learning: when we learn something, we attach cues to it and file like-cued material together.

In effect, what we learn early in life becomes the neural groundwork upon which all future learning is built and stored; it lasts a lifetime. It's difficult to overestimate the role good enough parenting plays in our lives. According to the **Adverse Childhood Experience (ACE)** test developed by Dr. Vincent Felitti, adverse childhood experiences are "still the best predictor of health spending, health utilization, for smoking, alcoholism, and substance abuse."[4]

THE STORY OF "ME"

Parents teach us who we are or, more precisely, they teach us how *they* see who we are. Just as there is no such thing as color or tone, smells, or tastes outside the brain,[5] "we are not born with an 'I' or a 'me'"…notions like these are both "social in origin and social in operation."[6] Somewhere around 18 to 24 months of age, babies start to distinguish between what's "me" and what's "other," and from that point forward, the idea of "me" takes form. The foundation of "me" comes from what other people tell us about ourselves – beginning with our name. Family members provide the initial building blocks; then extended family, teachers, coaches, babysitters, friends, and friends' parents. **Episodic memory**, the memory of personal life events, adds content and context to the idea of "me" and molds it into a coherent story. The story of "me" is the blueprint for our sense of self; it grows into our sense of self over time. "Me" and "I" become the principles around which we organize life: we create a unified perception of the world with "me" as the center and at the controls.

Of course, "me" is nothing more than a series of thoughts; synapses that have fired together so often that through long-term potentiation (LTP), they've wired together and become ever-present on a moment-to-moment basis. To the infant/child, parents add their experience of having children; their feelings about their own childhoods and projections of who this child is and could be. Through our ego structures, we internalize our parents' experience of us.[7] As Dr. Cozolino states, "me" emerges from relationships and exists as multiple layers of neural processing. It develops from the bottom-up[8]:

- From bodily awareness, to a subjective inner milieu;
- From sensory and motor processing, to a personal sense of agency;
- From emotional and cognitive operations, to arousal patterns, beliefs, intentions, and self-control;
- From social customs and traditions, to inferences, principles, judgments, and morals.

The **prefrontal cortex** (PFC) and the parietal cortex both play an important role in **self-recognition**.[9] In terms of **self-reflection**, the **medial prefrontal cortex** (MPFC) is the crucial player: it's said the medial prefrontal cortex gives us our sense of "I."[10] The brain, mind, and senses work together to create the one reality we live in and through: we take in some things and ignore others, constructing our perception which automatically includes our wants, drives, and dreams. Our unique neural architecture reflects our dominant **mind patterns**, memories, representations, acquired knowledge to date, implicit emotional experiences, and maps of and theories about how other people's minds work. Life provides the content of "me," out of which we form an identity. Our identity is what we hold onto; it's built on what we know.

Some authors believe that "me" is erected atop a deep layer of pain, one we spend our entire lives trying to avoid.[11] They suggest that personhood is one of the ways we attempt to avoid the pain, along with the reality we create. We work hard to keep things nailed down and under control, making everything appear "fixed, permanent, and lasting."[12] These "fixed" points represent our known truths; the ideas and beliefs we defend. The fact that "me" comes with a feeling of **agency** only reinforces the illusion. As a result, "me" feels so real that when one of our fixed points is threatened, it can feel as though our very existence is at stake. These authors suggest that in an effort to avoid the pain, we become a "self" living an all-absorbing life. We "consistently overestimate" the control we have over our lives, "while underestimating the role of chance, unconscious influences, and outside forces."[13]

Once "me" takes root, *everything becomes a projection of it*: "All elements of biology...and all past experience, whether remembered or long forgotten, affect the processing of incoming information."[14] Everything we attend to, perceive, infer, believe, experience, and do is a projection of "me"; it's created for, constructed by, and a reflection of "me." If you feel someone has been unfair, it's a projection of you. If you describe yourself as "doing the right thing," it's your projection about what you're doing and what constitutes "right." If you struggle with your partner's laziness, your boss' insensitivity, your best friend's infidelity, it's all a projection of you. Generally speaking, our projections just happen, although **conscious awareness** can intercede and force us to reflect on them. But as Dr. Cozolino points out, evolution has not yet seen the utility of investing heavily in the neural circuitry of self-awareness.[15] Perhaps this is so because conscious awareness creates doubt and uncertainty.

The story of "me" is an "ongoing personal narrative constructed by the very mind that is examining itself."[16] Every subsequent experience has to fit or *be made to fit* into that storyline. We repeat our storylines continually because they solidify our identity and reinforce who we think we are. Fortunately, the brain and mind come equipped with a built-in narrator to commentate our life's story.

THE LEFT HEMISPHERE INTERPRETER

The **left hemisphere interpreter** is the voice you hear in your head right now. Coined by Dr. Michael Gazzaniga from his research with **split-brain patients** (patients in whom the fibers of the corpus callosum have been severed), the left hemisphere interpreter is a **left-lateralized** system (it depends more on the left hemisphere of the brain) that is *driven to find explanations and interpret events.*[17] By providing an explanation, the interpreter allows the brain to cope with a situation in the event it should happen again. While the *right* hemisphere provides an accurate account of events, the left hemisphere takes more creative license: it interprets, predicts, and makes inferences based on what's presented.[18] The interpreter feeds the part of us that needs to know; irrespective of the circumstances, it assumes there *is* an explanation[19] and sets out to find it…or devise it, if necessary.

The left hemisphere interpreter narrates the story of "me." As the speaker of the mind, it is "intimately bound up with the first-person observational perspective."[20] It picks up where our family, friends, and others leave off, and explains our experience and behavior *to us*. The interpreter begins when we're very young, often with talking to our self. It is a self-soothing and descriptive tool: while the right hemisphere recognizes objects, the left uses language to make sense out of both, the right hemisphere and cognitive contents.[21] The left hemisphere interpreter is a silent voice that uses language to make sense of the world. It bolsters the brain's conservative nature and plays a supporting role when it comes to the three "goals" the mind and the brain seek to accomplish:

#1. To help make sense of the world, the left hemisphere interpreter:
- Narrates why things happen
- Is an inner voice that replays experience, ruminates, and stews[22]
- Reflexively confabulates -- it fills in memory gaps with guesses[23]
- Organizes experience and works to present the self in the best possible light.

#2. To help maintain a coherent sense of self, the left hemisphere interpreter:
- Integrates new information into our established storyline
- Constructs theories to perceive information as a "comprehensive whole"[24]
- Self-reports on the contents of consciousness[25]
- Engages in a form of story-telling, providing a narrative, not an analysis[26]
- Twists facts and dismisses contradictory evidence[27]

#3. To help sustain some semblance of control, the left hemisphere interpreter:
- Looks for patterns, tries to predict, and seeks order in chaos[28]
- Decides quickly based on little or sketchy information
- Avoids uncertainty, and is especially active when the world feels out of control[29]
- Uses causal attributions and makes causal inferences to feel in control

Gazzaniga's left hemisphere interpreter and the story of "me" bear a striking resemblance to the work of Don Miguel Ruiz.[30] According to Ruiz, we are not born with a voice in our heads. As we learn to talk, the voice is programmed into us and gradually takes over. Ruiz refers to the voice as the voice of knowledge. It is the sum of everything we've learned, a great deal of which is not true. It builds on shaming from childhood and early messages that tell us we can never measure up. Slowly it fashions itself into our sense of knowing, a point of view that we project outward and try to impose on the world. In effect, the voice is telling us a story that "qualifies, justifies, and explains" what we perceive, creating a world that is no longer real, "it is a virtual world." [31] Ruiz states, "Your story is your reality – a virtual reality that is only true for you, the one who creates it…the structure of our knowledge makes us feel safe."[32] To step outside that virtual reality, we must first notice the voice and the role it plays.

A FILTER OF PAST

All of this becomes *past*. Everything our parents teach; every early learning experience, pattern, association, representation, and inference; every explanation advanced by the left hemisphere interpreter and every chapter in the story of "me" becomes past. The past influences and biases our experience of both the present and the future. We *are* our past: everything that happens to us is processed through the past. The brain makes this so. When we see an image, for example, what we see is a combination of the image and our past experience: "our vision system evolved to make sense" of what we see…which is to say, "We see what we expect."[33] Similarly, our re-actions and everyday behaviors are a reflection of the past: we don't consciously scan the archives every time we're faced with a situation -- we automatically resort to what's worked in the past. One author calls this **"synaptic shadows**," the ways in which the past casts its influence over our present and future experience.[34]

In effect, every experience becomes part of a large accumulating filter through which, thereafter, everything passes. While the filter continues to evolve, it is heavily weighted towards our early learning: our underlying sense of being "okay, capable, and loved," our sense of safety, and the implicit learning we acquired unconsciously. And the filter casts a wide shadow. It influences:

- What we pay attention to;
- How we file memories and whether we can recall what we've learned;
- What's important to us and why;
- Our self-definition and how we see our future;
- What we define as a threat and an opportunity;
- The assumptions, perspectives, and judgments we form;
- How we appraise incoming information, make decisions, and analyze data;
- How we relate to others, the attachments we form, and the depth/quality of our relationships; and,
- The expectations we have of others and of the world generally.

The past influences everything we say, see, and do: we approach new situations based on their similarity to past ones; make predictions based on what's happened in the past; decide the relative importance of something based on what's been important in the past. Rather than "perceive the world anew every moment…we automatically reconstruct the reality we perceive from models...stored in our memories."[35] Hence the world we experience is "doubly removed" from reality: first, by our "perceptual apparatus" that "sample and represent" only "certain selected features," and second, by our memory.[36] We even *live* in the past: because it takes time for information to be processed through our filter of past, we live about 500 milliseconds after the fact,[37] despite our experience of living in the present.

Our filter gets incorporated into the story of "me" as narrated by the interpreter; it becomes *my* past. It embodies the long list of factors that vary from person to person: fetal experience and genetic differences; thresholds of neural excitability; the labyrinth of neural connections in the brain and the strength of synaptic connections; one's sensitivity to the environment, emotional homeostasis, and arousal set-points; the capacity for long-term potentiation (LTP) and the brain's capacity for plasticity; the experience of one's subjective inner milieu; and the "continual emergence of dynamic structures...molded by connectivity and subtly modified by external input and internal state."[38] These differences among individuals are even more pronounced in the recently evolved areas of the brain. And they manifest socially: individual differences are reinforced by social distinctions. They strengthen our sense of "otherness" and foster dualistic thinking. As we grow, it becomes exceedingly difficult to get outside our own box and beyond the confines of the filter...until at some point, we stop trying.

From this section on individuality comes a critical insight from the **wisdom** of the brain: **we cannot know the experience of another person**. It's simply not possible. What it feels like to be another person, what he/she perceives or thinks, what he/she needs – *we cannot know*. The fact that a full 70% of the brain is developed post-birth[39] means that the bulk of our brain's connections depend on what happens to us. Given that, there is no way we can know the experience of another. This means that as helping professionals, we cannot make specific recommendations. Recommendations are like hope: having hope is fine, but when we hope for something in particular, it becomes a set-up; it's destined to disappoint.

Section V

The Brain and How We Accumulate Knowledge

18

How the Brain Learns

The goal of **learning** is for selected items to reach long-term memory. The goal of **memory** is to retrieve and apply those items. During learning, information is temporarily held in working memory where it is rehearsed, linked with previous learning, coded for long-term storage, and then moved to memory storage sites. To retrieve and use information, an internal or external cue triggers a search of long-term memory. The item is reassembled in working memory and if warranted, integrated with new information. It is then reconsolidated and returned to long-term memory storage. Under especially tense or stressful conditions, people can lose their ability to recall or sequence events. Sometimes, memory fails entirely. Helping professionals need to understand why this happens, but also how and why people remember what they do…and why memory falters even under the best of circumstances.

IT *CAN* BE TAUGHT
Learning is possible because of **neuroplasticity**; it is a measure of our adaptability. The fact that we can learn means we don't need to rely on instinct to tell us what to do: we can create possibilities, make choices, and figure out new techniques to increase survival. Learning weaves together many of the topics we've been discussing:

- Learning depends on motivation and desire; it's derived from our basic **seeking systems**. It begins with **patterns** of language, learning, and memory that are laid down in childhood. Research has found that learning happens best when the learner is curious and motivated; when the material is somewhat challenging, exciting, and interesting; and when the setting is safe and supportive.

- Learning builds on experience: "Genetic and developmental processes specify the connections among neurons...But they do not specify the strength of those connections."[1] That depends on experience, on learning. We are born with an overabundance of **synapses**, of "potential connections" that "represent all the possible worlds we might find ourselves in."[2] Learning "selects among a large repertoire of preexisting connections and alters the strength of a subset of those connections."[3]
- Learning creates meaning. Meaning is not innate; we have to learn what has meaning to us. Most of our "cognitive belief system" is learned,[4] passed down through narratives, laws, and traditions. Similarly, the majority of "our thoughts, emotions, and behaviors are shaped by meaning systems of our own creation."[5] Because of this, "conflict among meaning systems is the source of some of our most profound hatreds."[6] A threat to one's personal meaning system can feel like a threat to one's very sense of self.
- Learning relates back to the human need for certainty. While practical knowledge has built-in utility, the *feeling of knowing* may be the brain's reward for acquiring abstract knowledge: "from hunches and gut feelings to faith, belief, and profound certainty"...the feeling of knowing has evolved as "abstract thought's subliminal cheerleader."[7] That feeling is present whether or not what's learned is factual.[8]
- Learning creates past. It's likely that the memories we acquire early in life form the deeply consolidated trunk circuits upon which long-term memory is built, including memories with conscious awareness called **explicit memory** and unconscious memories called **implicit memory**. Together they form the filter through which all input is processed.

In order to learn, certain conditions have to be met. First, there has to be desire or motivation on the part of the learner. The best motivation is intrinsic because the learner has his/her own reason to learn the material. Second, the neural pathways have to be in place because learning strengthens *existing* pathways. Third, there has to be a stimulus to prompt learning, the material can neither bore nor overwhelm the learner, and the environment must be physically safe and emotionally secure.[9]

Learning involves the acquisition of knowledge. To retain it, the new material has to be linked with older, established knowledge and associations need to be formed. The more associations we have, the more we can make, and the more we can generalize, infer, deduce, and create from those associations. New material is then stored with items of a similar nature. Through **transfer**, we can take a strategy from one situation and apply it to another, similar one. Not all memory is learned and a great deal of it is not remembered. And as we shall see, knowledge is not the same as understanding; memory requires more than simply learning the material; recognition is not the same as retrieval; and, application requires more than just remembering…

THE LEARNING BRAIN

When we learn something new,[10] sensory input goes to the **sensory register or memory** where, via the thalamus, information is screened against past experience to assess its relative survival value and importance to the individual. Information is retained in sensory memory for an extremely short period of time, from hundreds of milliseconds up to a second. If the sensory input is important enough, it then moves to cortical sensory processing areas and to **short-term or immediate memory** where it is held briefly to decide what to do with it, such as rehearsal. In terms of sensory data, a processing hierarchy is in place: input that involves a threat to survival is given top priority, followed closely by that which is emotionally salient, and finally, data for new learning.[11] Input that involves conscious processing is sent to working memory, while information processed outside conscious awareness involves implicit memory networks. Only a fraction of what comes into short-term memory ever makes it to long-term memory.[12]

Working memory is another conception of short-term memory as a temporary site where information can be considered and manipulated for eventual long-term storage. It is also a mechanism by which information in long-term memory can be summoned into consciousness for present use, such as recalling facts about previous events or places. Working memory is where the decision is made whether or not to save new information. Information of survival or emotional value is quickly moved to long-term storage. Beyond that, the decision to transfer an item to long-term storage hinges on two issues[13]:

- First, whether the new material *makes* sense; and,
- Second, whether it has *meaning* for the learner.

The first issue has to do with whether the new material fits with the learner's past experience, with "what the learner knows about how the world works."[14] The second pertains to whether the material is relevant to the learner. If the learner cannot understand what's being taught or why it matters, chances are it will not be preserved in long-term memory. Moreover, between the two, *meaning is more important*.[15] It's not enough to know the material -- learning requires having thought about why it's important and how it's relevant to "me."

It is through the process of repetition in working memory that the learner makes sense of the material and assigns meaning to it.[16] Repetition for complex cognitive learning is called **rehearsal**; for motor skills, it's called **practice**. Both rehearsal and practice involve frequent repetition. Through the practice of motor skills, the sequence of movements grows automatic and unconscious until it becomes a fixture in **procedural memory**. Additional practice refines and varies the basic skill set. For complex learning, the more time spent rehearsing the material, the more likely the learner is to make sense and find meaning in it. Abstract learning in particular has to be "thoroughly and deeply processed…associating it meaningfully and systematically with knowledge already well-established in memory."[17]   It requires a special kind of repetition -- called **elaborate rehearsal**[18] -- that allows for review, practice, and rest; and takes a variety of forms including paraphrasing, note taking, questioning, summarizing, and application, all aimed at improving retention.[19]

After working memory, the next stop is **long-term memory** storage. In select instances, information is retained in working memory where it's worked on outside conscious awareness; working memory "chews" on the problem until a solution is reached that either enters conscious awareness or is filed away. In long-term storage, memories are stored as patterns of connections. New learning is "filed" based on how the brain creates meaning and makes sense of the information. Memory storage itself is a skill: how well information is encoded and stored determines how easily it can be retrieved. As a learned skill, there are techniques that improve the speed and accuracy of memory retrieval.[20]

You may also be interested to know that:

- The brain learns throughout life: the more diverse our interests, novel our experiences, eclectic our relationships, and the more opportunities we create, the more resources we have at our disposal when we need them;
- Rather than perfect, practice makes *permanent*.[21]   To *improve* performance takes motivation; it requires that the knowledge we need is accessible, comprehensive, and applied correctly;
- Both hemispheres in the brain work together in learning. For example, the left side of the brain may know a word, but that's not enough. When the right hemisphere chimes in, the context, meaning, and application of a word is known as well;
- Learning and retention vary by teaching method: learners tend to remember what's taught first and last; multiple short periods of learning are preferable to one long period; visual aids enhance learning and practical application enhances memory. Having students teach one another is also very effective[22]; and,

- While specific areas of the brain specialize in learning spoken language, there are no such areas devoted to reading.[23] With respect to video games and computers, these affect learning by decreasing the time available and "willingness" to pause and reflect, and think deeply.[24]

Learning codifies events and outcomes from the past. The past serves as the basis for whether new information is retained, where it's stored, and how it's encoded; for judging whether our performance has improved and deciding if further practice is worth the effort. The past affects nearly every aspect of how the brain learns.

LEARNING AT THE CELLULAR LEVEL

Whenever something new is learned, it forms a new pattern of connections: the **dendrites** of neurons receive a signal; if the change is positive and sufficient to meet a given threshold, a spike or **action potential** is propagated along the **axon**; to progress to subsequent neurons, a **neurotransmitter** must be released which diffuses across the synapse and binds with receptors of fellow neurons; and then the process begins again. With rehearsal and practice, the tendency for neurons to fire together in the future increases, thus forming new patterns of connections. Ultimately, they wire together such that when one neuron in the pattern fires, they all fire, and the material is effectively *learned*.

Learning depends on the brain's **plasticity**. It requires synaptic plasticity and in some cases, **synaptogenesis**, as well as **long-term potentiation** (See Chapter 5). Synaptic plasticity and synaptogenesis effect a **functional** change in the brain; they increase the strength of connections between neurons. Long-term potentiation (LTP) involves the repetitive and simultaneous firings at multiple sites along the pathway. It causes the dendrites to become more sensitive to stimulation and makes the pattern more likely of firing together in the future.

NMDA receptors are known to play an essential role in long-term potentiation and the changes in synaptic strength which underlie learning and memory. NMDA stands for N-methyl-D-aspartate and its receptors are found on the spines of postsynaptic neurons involved in long-term potentiation,[25] especially in the hippocampus.[26] The activation of NMDA receptors can increase the "potency" of the synapse, causing synapses to grow stronger.[27] Interestingly, long-term potentiation is only *thought* to be the way the brain learns and remembers. It is an "appealing explanation" because of its **specificity** -- only those synapses that convey information relevant to the memory are strengthened; its **cooperativity** -- only those events that meet a certain threshold are remembered; and its **associativity** -- events that might otherwise be insignificant, gain in importance when paired with, or are part of, something worth remembering.[28]

Learning in the hippocampus, which is capable of **neurogenesis** – the creation of new neurons, is unique in that it first involves changes in synaptic strength (synaptic plasticity), then the growth of new neural connections, and finally, new neurons. Synaptic plasticity and synaptogenesis, however, happen far more frequently than neurogenesis and are the learning methods of choice at the cellular level of the brain. Once synapses are strengthened, the connections still need to be used and maintained. Remembering requires more than simply learning the material -- rehearsal and practice need to be followed by application, use, and re-use. Absent that, we lose the skill because we lose the structural changes in the brain that support it.

ENTIRELY PLASTIC?

Plasticity demonstrates the brain's **self-organizing** and **emergent** properties: there is no pre-set plan for the brain -- we spend a lifetime adding, deleting, and rewiring its circuitry. Moreover, at least for humans, evolution has placed greater emphasis on retaining *existing* knowledge and building onto it, than on generating new neurons,[29] a fact which attests to the brain's **conservative** nature. But just how plastic *is* the brain?

Neuroplasticity does not mean that every undesirable aspect of ourselves can be changed; perhaps some are too deep or for some other reason, impenetrable. While many books on the market mention the incredible spatial plasticity of London cab drivers or the expanded somatosensory cortex of expert violinists, the truth about plasticity lies somewhere between exceedingly plastic and altogether rigid. The **wisdom** of the brain tells us **that we can learn what we need to know**... and life has a way of repeating what we resist learning. The promise of plasticity needs to be tempered with compassion, patience, and fairness.

Once established, **unlearning** involves either replacing or suppressing/inhibiting learned material. When a motor skill has been learned incorrectly, successful relearning depends on the learner's age (the younger, the easier it is to relearn); the length of time the skill was "practiced incorrectly"; and the learner's motivation.[30] Learning that happens early in life, and especially emotional learning, is incredibly hard to unlearn precisely because it is tied to survival. Like learning, unlearning depends on the brain's plasticity and there are several options. Just as repetition increases the strength of synaptic connections, lack of use will diminish those connections. Rigid self-regulation is another option, although it comes at a price which includes the shame and feelings of self-hate that often accompany it. Perhaps the best approach is to replace an old thought or habit with a new one which, at least initially, requires bringing the issue into conscious awareness to avoid reverting to the ever-present Path of least effort.

Recommended Reading
How the Brain Learns, by Dr. David Sousa

19

How the Brain Remembers

Memory stores the past for future use. A memory can be a recollection of a past event or fact, a learned motor skill, or an unconscious influence. To remember, an internal or external cue triggers the brain to search its files and retrieve the information from storage sites in the **neocortex**. The information is then reassembled in **working memory** and readied for application to the current situation. Generally speaking, the key to memory is its retrieve-ability – retrieve-ability depends on how the brain makes sense of, and finds meaning in, the material, and links it to the existing knowledge base.

The **medial temporal lobe, the prefrontal cortex,** and **the temporal cortex** comprise the biological memory system.[1] The medial temporal lobe includes the amygdala, hippocampus, and several other areas.[2] The **hippocampus** is heavily involved in the *creation* of *explicit* memory, moving it from short- to long-term memory. Memory includes:

- **Encoding** refers to the consolidation of new material and assigning it a code or reference marker so it can be located in the future. Interestingly, encoding is more of a left hemisphere activity while **retrieval** depends more on the right,[3] and different types of memory can be encoded in the same cell. Encoding takes time – an estimated 70 to 90 percent of new learning is lost within the first 18 to 24 hours.[4]

- **Storage** involves the creation of a record or file of the new material. Explicit memories are stored in pieces, in multiple storage sites, and in networks containing similar items. They are stored as "distributive representations" in the area of the neocortex that "originally encoded" the information and other regions associated with that information.[5]
- **Retrieval** involves locating information from its long-term storage sites and reassembling it in working memory so it can be used. While information is stored by similarity, it is *retrieved by differences* -- by how it "shines out" from other material from the perspective of the learner.[6] Recognition is not the same as recall: **recognition** involves matching an "outside stimulus" with information stored in long-term storage, such as in a multiple choice test. **Recall** is harder: it requires using a clue or a hint to search, locate, and retrieve the desired information, and then reassembling it in working memory.[7]
- **Consolidation** refers to the jelling or solidifying of a memory. Consolidation is part of the encoding process and happens best when the brain is not occupied, as during sleep. Researchers believe that during sleep, memory bits from various storage sites are reassembled and replayed, firming up the linkages between them.[8] Once consolidated, when a memory is recalled, new or additional information may need to be added to the original memory. This process of **reconsolidation** also happens best in a quiet, calm setting or un-aroused state.

New memory networks can be thought of as being "stacked" on top of old ones, building from elementary information to more advanced and abstract knowledge; while organized hierarchically, there is a considerable degree of interaction between the layers.[9] As such, despite our best intentions, memory is fallible: the retrieval of a memory involves the "reactivation of patterns similar to, but never identical with, the initial encoding."[10] It's been said that each time we recall a consolidated memory into working memory, we relearn it.[11]

Memories are "traces of influence from the past" that continue on into the present.[12] While there is a standard structure around which human memory is organized,[13] the content of each person's memory is unique, which makes memory a major contributor to our individuality. Alongside parenting styles, early emotional learning, and life experience, memory is another reason why we behave as we do.

TYPES OF MEMORY AND THE BRAIN SYSTEMS THAT SUPPORT THEM

The organization of human memory includes **short- and long-term memory**. Short-term memory includes **sensory memory** and **working memory**. Long-term memory is divided into **explicit memory** (also known as declarative memory), and **implicit memory** (also known as non-declarative memory). **Explicit memory** can be either episodic (auto-biographical) or semantic (factual). **Implicit memory** generally includes procedural or skills memory, perceptual representation and priming, classical conditioning, and non-associative learning in the form of habituation and sensitization.[14] The various types of memory are described below:

Sensory Memory

- Lasts from hundreds of milliseconds to a second
- Provides high capacity temporary storage
- Includes **echoic** (auditory) and **iconic** (visual) memory
- Transacts outside conscious awareness
- Only a fraction reaches long-term storage

Working Memory

- Timeframe depends on the task but is short-term in nature
- Includes a **phonological loop** (acoustical) and a **visual sketchpad**
- Used as temporary storage to manipulate and work on information
- For humans, input is in chunks of 4-7 items

- Input can come from sensory or long-term memory
- Is supported by the dorsolateral prefrontal cortex and hippocampus

Long-Term/ Explicit and Implicit Memory

- Duration of days to years
- Hippocampal system is key to creation of *explicit* memory
- Explicit memory can be expressed through conscious awareness
- Subcortical/amygdala are key to the formation of *implicit* memory
- Implicit memory can at times be expressed, but the source is unknown
- Long-term explicit memories are stored in distributive networks in the neocortex
- Hippocampus allows us to draw inferences from past situations and apply them to new situations

Explicit/Semantic Memory (fact-based)

- Semantic memory is most easily recalled under conditions similar to those in which it was learned
- **Left frontal cortex** is involved in encoding and retrieval of semantic memory[15]
- Semantic memory is stored in the **temporal cortex**[16]

Explicit/Episodic Memory (autobiographical)

- Episodic memory includes both content and context
- Episodic memory is consolidated early in the sleep cycle when cortisol levels are low
- Medial temporal lobe is thought to bind information together into an episode[17]
- Left frontal cortex is involved in encoding, **right frontal cortex** in retrieval[18]

- Episodic memory is stored in the temporal cortex[19]
- Episodic memory is the only backward-looking memory

Implicit/Procedural Memory

- Includes learned and practiced motor skills that become automatic over time
- Includes certain repetitive cognitive skills[20]

WORKING MEMORY – "THE BLACKBOARD OF THE MIND"

Without memory, we couldn't encounter a situation, recall one similar to it, and come up with a ready-made solution. We would have to relearn the same thing every time we needed it. We couldn't recognize a face, keep track of what we're doing, or string together a series of thoughts, let alone a conversation. One reason we can do these things is because of **working memory**. Working memory has been called the "blackboard of the mind."[21] It's a temporary repository for representations of information that are relevant to the task at hand. Working memory is where information is retained during the learning process; it's where memories are reassembled after retrieval, and *where you're reading this book right now*, linking one sentence to the next.

Working memory has a capacity of between four to seven (4-7) items. To expand this, humans group like items together under a single category in a process called **chunking**. Nevertheless, the capacity of working memory for humans appears to be decreasing for reasons that are not yet known.[22] The model of working memory typically includes a phonological loop and a visual sketchpad, and often a **central executive**. The central executive is thought to coordinate the other two systems, and is especially active during cognitively demanding tasks. The central executive is believed to keep working memory continuously updated, allowing us to juggle multiple items, compare and contrast among items, and reason, judge, and predict.

As opposed to residing in a particular place in the brain, working memory is a distributive system or process -- the areas in the brain that are activated depend on the task at hand. Working memory is supported by the **dorsolateral prefrontal cortex**, the hippocampus, and other areas. General associations are fed into working memory from the interaction among these areas and for this reason, it can be thought of as a "conversation" between new learning and the past or established knowledge.[23] Working memory is essential to how we organize thoughts, ideas, decisions, plans, and actions. It makes **non-stimulus driven behavior** possible: because of the projections between the hippocampus and the cortex, humans are able to "recreate" "a version" of the pattern of neural connections that occurred at the time of the original experience[24]...we literally bring something to mind *absent* the originating stimulus. While working memory can be overwhelmed by stress and/or threat, we rely on it nearly every moment of the day.

EXPLICIT AND IMPLICIT MEMORY

Explicit memory is usually defined as "the conscious recall of people, places, objects, facts, and events."[25] It is divided into semantic and episodic memory. In contrast, **implicit memory** is "a collection of processes involving several different brain systems."[26] It includes **procedural memory** which is supported by the **striatum** and learned through the **cerebellum**. Emotionally salient memories (supported by the **amygdala**), beliefs, unconscious associations, early emotional learning such as attachment schema, etc. are also implicit. Implicit memory is generally considered **unconscious** because it is largely inaccessible to conscious awareness and its influence is automatic. While we can train ourselves to consciously notice an implicit influence – such as every time we have a discriminatory thought -- we'll never know for certain where those thoughts came from. A great deal of learning relies on *both* explicit and implicit memory: by practicing one's golf swing consciously and deliberately, the explicit memory of the swing gradually becomes implicit, automatic, and unconscious.[27] The then-*procedural* memory of the golf swing is perfected, updated, and refined over time through practice.

Figure I compares explicit and implicit memory:

Figure I
Types of Long-Term Memory

Explicit Memory (Declarative)	Implicit Memory (Non-declarative)
• Occurs later in development • Is referred to as conscious memory • Can be explained, declared, or demonstrated[28] • Can falter under stress • Hippocampus can be shut down temporarily from alcohol, rage, other states of high emotion[29] • Includes episodic and autobiographical narratives that become part of our personal story • Depends on language, and may require a sense of self -- an "I," the teller of the tale • Focuses on the issues, matters, objects, etc. that we attend to • Includes integrated (known) biases, as well as social rules, expectations, customs, etc. • Is full of errors	• Occurs early in development • Has an automatic quality: "nonverbal, non-conscious and embedded in the body"[30] • Is the only kind of memory encoded during the first 18 months of life • Not a single system but a collection of systems • Can be misinterpreted as intuition or insight[31] • Has no conscious access to the source, is not integrated by the hippocampus • Shapes emotional re-actions, subjective feelings, perceptual/unconscious biases • Includes fear conditioned memories, early attachment and relationship schemas • Captures information that occurs outside conscious awareness[32] • Is difficult to extinguish

Explicit memory includes **episodic memory** and **semantic memory**. Episodic memory can be described as the "dated recollection of personal experiences."[33] It includes our conscious awareness of past events as well as our personal autobiographical memories. In contrast, **semantic memory** houses factual knowledge: it is "general information" that is "stored and undated."[34] Generally, there are no events and episodes associated with semantic memories. Some authors distinguish between episodic and **autobiographical memory** with the former being more trivial and short-term, while the latter stores material of greater significance to the individual concerned.[35] Autobiographical memory in particular, can be thought of as a tug-of-war between maintaining the accuracy of the memory and its coherence with our personal goals and beliefs.[36] More often than not, over time, coherence wins out.

Other interesting facts about episodic and semantic memory include:

- Semantic and episodic memory provide a base of past knowledge to draw on and apply to current situations, as well as to new or novel ones;
- Episodic memory is thought to be a recently evolved form of memory. It is a "late-developing, early-deteriorating past-oriented memory system"[37];
- Semantic memory is organized into categories of information, although the categories are not rigidly defined but are "more fuzzy" in nature[38]; and,
- Episodic memory is a "*constructive* process," not a reproductive one.[39] The parts of ourselves we prefer not to remember in a particular way can be revisited and revised.

Procedural memory is part of implicit memory and has to do with producing purposeful action, both cognitive and motor skills. Procedural memory skills are sometimes called "automatized" behaviors because they are learned rather than "instinctive." Examples of procedural memory-based motor skills include riding a bike, typing, or fluidly dancing the tango; procedural memory-based cognitive skills include reading, figuring out a repeatable way to solve a problem, memorizing a song, etc.[40] The whereabouts of procedural memory in the brain depends on the nature of the task. During the early stages of learning, the **premotor cortex**, working memory, and the cerebellum are all actively involved in acquiring a new motor skill. Learning requires repetition in the form of practice, and improvement through feedback such that the skill can be continuously refined. It takes several hours for whatever is learned during a practice session to consolidate in the cerebellum.[41] Once mastered, the premotor cortex and working memory are no longer needed for execution; the cerebellum takes over and performs the skill through procedural memory.[42]

There is a clear **disassociation** or distinction between the brain systems that support explicit memory and implicit memory.[43] Procedural memory provides a perfect illustration of this: if you consciously think about a procedural memory, you will *decrease* its accuracy. Try, for example, consciously writing your signature and see what happens. Thankfully, execution doesn't require the multiple areas of the brain that store explicit memories.

Beliefs are part of implicit memory – we are often not aware of their source. Two well-known authors in memory research, Howard Eichenbaum and J. Alexander Bodkin, argue that beliefs are *not* the same as knowledge: "Knowledge is a disposition to behave" that is corrected and updated by experience; belief is also a disposition to behave but it is "resistant to correction by experience."[44] They theorize that beliefs and knowledge may reflect distinct modes of processing in the brain and use memory differently. In particular, beliefs may be supported by pathways in the amygdala and the **neostriatum**.[45]

The **wisdom** of the brain tells us that **how we learn affects what we learn**. These authors suggest that when we learn a new topical area, we begin with a **knowledge-driven approach**: we gather information, take note of its origins, form associations around ideas, etc. Over time, as we accrue additional information, the original sources of those associations become lost. After many iterations of this, the end result is "a knowledge base that has all the properties…[of] cortical-hippocampal interactions": it's a "large organized network of associations" that is flexible, supports "inferential memory expression," and is updated through new experiences.[46] As the process continues however, and new experiences "largely confirm the existing framework and add less and less new information," we shift from a knowledge-driven approach to a **belief-driven approach**.[47] We begin to assume that *all* new information will fit within the existing framework, and when it doesn't, we either ignore it or find a reason to dismiss or discredit it. Only when we start to accumulate *too much* information that doesn't fit are we forced to switch back to a knowledge-driven approach.[48]

WHEN MEMORY FAILS

Many people still believe a permanent record is created in memory for each experience and forgetting is simply a problem of retrieval.[49] However memory is fallible for many reasons; beginning with the fact that there are multiple storage sites, it requires reassembly, and the process is constructive rather than reproductive. In addition, memory tends to be mood-based: when we feel happy, we're apt to recall pleasant memories.[50] Our expectations can distort memories,[51] as the brain tries to anticipate what will happen next. Memory is also sensitive to context, and complex memories draw on more networks, making them harder to accurately reassemble.[52] Daniel Schacter, an icon in the field of memory research, suggests there are seven "sins of memory"[53]:

1. Facts and events often become less accessible to memory over time;
2. Forgetting is often the result of not paying enough attention when the information was initially learned;
3. Memory can be blocked for a whole host of reasons;
4. Memory is characterized by the "sin of commission" -- certain parts of a memory may be intact but the time, place, or person can be misattributed;
5. Memory is prone to suggestibility: the suggestions of others, a leading question, etc. can influence what one remembers;

6. Preexisting knowledge, beliefs, and feelings can bias memory, and, due to "consistency bias," people tend to "exaggerate the consistency" between past and present beliefs and attitudes[54]; and,
7. Memory suffers from the fact that it's very hard to forget a memory you want to forget.

Autobiographical memory is particularly malleable, but memory in general is vulnerable to unconscious fabrication, to a process called **confabulation**. When a memory is retrieved, if the information is incomplete or a piece is missing, long-term memory will "unconsciously fabricate" the missing piece by "selecting the next closest item" that can be recalled.[55] The less remembered, the more the brain has to fill in. The memory is then reconsolidated, including the fabricated pieces, which then becomes what we believe to be *the* true picture of what happened.[56] While not a deliberate process, confabulation illustrates the lengths to which the brain will go to avoid admitting we don't know. We could easily question whether *any* memory is ever really accurate.

Finally, in terms of memory and aging, the adult brain learns throughout life but it has to have something to learn: the absence of new and challenging information and experiences means we increasingly rely on the same old, repeatedly reconsolidated circuits. Inevitably, our perspective narrows. According to an article in the New York Times, *The Secret of the Grown-Up Brain*, adults learn best when confronted with ideas that are different from their own.[57] Absent this, we default to the path of least effort: we recycle what we have…as mixed up as it might be.

A
Closing Thought

A Neuroscience Question: How does memory support our sense of self?

Memory is an accumulation of what we know. While the self is built from the bottom up, our *identity* is heavily tied to what we know,[58] making the contents of memory a defining feature of the self. At some point, our sense of self separates from our biology and the self becomes an essentially mental construct.

Rich **episodic memory** is said to be among "the hallmarks of human consciousness," situating our sense of self in a particular place at a particular time.[59] Because of the extensive projections from the neocortex to the hippocampus, we're able to remember not only life's events but also the meaning we ascribe to them.[60] While we generally recall *the gist* of our everyday experiences more so than the precise details, according to the **multiple trace theory**, every time an episodic memory is retrieved from long-term storage, it is re-encoded, leading to widely distributed "multiple...traces" of the same event.[61] It's generally believed that episodic memory includes "systematic distortions" and many errors.[62] As the only backward-looking memory, episodic information takes the form of a narrative, a form of "subjective time travel"[63] in which the teller tries to present him or herself in the best possible light. It provides the content, context, and consistency for the evolving story of "me," that is narrated by the **left hemisphere interpreter**.

In terms of **belief systems**, the self has a vested interest in preserving what we know and avoiding new information that undermines our knowledge base. This is because beliefs allow us to feel as though the important issues in life are nailed down and under control, continuously reinforcing a *coherent* sense of self. The more something has become a belief, the more restrictive we are in interpreting new information, forcing it to "conform to the preexisting information."[64] We hold on to our beliefs even in the face of mounting evidence to the contrary, preserving what we know for as long as we can; forever, if possible.

Therefore, both explicit and implicit memory create and feed our sense of self, gradually drawing the mind away from the body and into the head, designing and redesigning the storyline that we call "life." It includes everything we know and believe, and all the stories we tell and the explanations we give; each of which reflects some version of the original experience…our *believed* truth.

Section VI

Conscious and Unconscious Influences

20

The Scope of the Unconscious

Everything manifests in consciousness and the unconscious – the effects of parenting, our emotions, knowledge, memories, perception, thoughts, etc. Some of our mental processes are conscious. Others can be *brought into* conscious awareness, and sometimes are. Most are neither part of, nor accessible to, conscious awareness -- they are ever-present but unconscious, and explain a great deal about why we behave as we do.

THE PRINCIPLES OF THE BRAIN

In an effort to summarize what's been discussed so far, we could say the brain operates according to according to five key principles:

1. Neurons connect with other neurons to form pathways of communication. Of all the possible pathways, the ones that survive are the ones we use the most.
2. The brain is integrative; it works together as a whole. Regions can specialize, but processes are distributed. The brain shapes massive quantities of information into an integrative whole.
3. The brain naturally looks for patterns. Patterns grow into categories; categories become generalizations; and beliefs, assumptions, and expectations flow from there.
4. The brain learns from experience. The neural architecture of the brain changes as a result of life experience.
5. The brain learns and remembers what has meaning; meaning refers to what's important and has value to "me."

We are now going to add a sixth principle: *the brain makes as much as possible automatic and unconscious.* The first five principles relate to the brain's embodied and emergent properties. This new principle speaks to the brain's conservative nature: because the brain is conservative, it seeks to "automatize" whatever it can and file it away, reserving the bulk of conscious attention for particularly difficult, novel, and demanding situations. Just as we don't know the source or accuracy of much of our memory, we are not aware of the extent to which the unconscious drives us. The fact is…we barely know ourselves.

THE WORK OF THE UNCONSCIOUS

Interest in the nature of consciousness and the unconscious dates back to the earliest spiritual pursuits and throughout history, philosophers and scientists alike have tried to understand it. Conventional wisdom views the unconscious as the lesser of the two states. Whereas consciousness is seen as deliberate, contemplative, the source of reason and rationality, the unconscious is often equated with negativity, uncontrolled impulses, and irrational behavior. Over time, this distinction has grown outdated as the age-old divisions between reason, conscious, and control on one hand, and emotion, unconscious, and uncontrolled on the other, are fading away.[1] So, too, is the timeworn idea that the cerebral cortex, and the prefrontal cortex in particular, participates in only conscious endeavors. Research has found that many activities once thought to be strictly conscious are now known to occur unconsciously as well, and that the cerebral cortex is involved in many different types of unconscious processes.[2] This, of course, challenges some of our cherished notions…such as free will, intentionality, and moral responsibility.

Despite its reputation, our ability to "automatize" a whole range of mental processes is "a gift of evolution," one that's "crucial to our survival as a species."[3] The sensory system sends the brain 11 *million* bits of information each second, but we can only manage between 16 to 50 bits per second.[4] That means that while consciousness is focused on whatever needs attention, the unconscious is attending to everything else. It does so by **routinizing** how we process information and then respond to it. As a result, we can greet someone, read an expression, make a decision, gesture in support, and spot a familiar face -- simultaneously, seamlessly, and without much thought.

No specific part of the brain houses "the unconscious"; nor is any region more or less involved, although implicit memory is known to depend on the **striatum**, the **cerebellum**, and the **amygdala**. Unconscious influences come from neural connections that have evolved over time through natural selection, genetics, and an individual's past experience. It's another way the brain makes sense of the environment, imposing order on chaos. It's estimated that a full "95% of our actions are unconsciously determined."[5] Put another way, 95% of what goes on in the brain -- of our "**cognitive function**" -- happens outside conscious awareness.[6] Although that does not mean that 95% of our behavior is unconscious, clearly most of what we say, think, do, and feel is shaped by unconscious processes: "The conscious you…is the smallest bit of what's transpiring in the brain… Our brains run mostly on autopilot, and the conscious mind has little access to the giant and mysterious factory that runs below it."[7]

By routinizing how we process and respond to information, the unconscious generates efficiencies and saves energy. It enables many of our thoughts and behaviors to be preprogrammed, pre-planned, and formulaic. The unconscious is part of our filter of past; it *is*, in effect, the path of least effort. It allows the brain to conserve what is known, "our pre-existing knowledge, beliefs, attitudes, and stereotypes."[8]

UNCONSCIOUS INFLUENCES

Unconscious processes affect *all* the ways we manage incoming information and our responses to that information. In fact, that's how the **filter of past** works. Through associations, causal links, and inferences, the brain shapes the past into maps, rules, scripts, and models. From those, we produce beliefs, judgments, and assumptions which in turn, become part of our ever-expanding, *unconscious* filter of past. While we can, with effort, become aware of some unconscious influences, many are not known and some are not *knowable*; they never come up for conscious review and revision. Unconscious influences include:

- <u>Constructed Perception</u>. To construct our perceptual experience, the brain takes sensory input and adds memories, states of mind, expectations, desires, etc., to fill in any gaps and smooth out the rough edges. Our perceptual systems are so automatic that the brain can readily incorporate smart technology as an added source of information, making it "part of itself" and unconscious to the user. There is, as a result, "no 'immaculate perception'… [it's] virtually always a blend of what we are sensing now and what we've learned previously."[9] Yet we believe what we perceive and forget that it's just "activities of the mind."[10]
- <u>Automatic Body Movements</u>. The body operates largely outside conscious awareness: the **autonomic nervous system** runs without conscious intervention; **reflexes** happen before we're consciously aware of them; and **procedural memory** is so deeply ingrained we don't need consciousness to execute most of our movements. Instead, we reserve consciousness for when something goes wrong, like we cut ourselves or feel ill.
- <u>Instincts</u>. Preprogrammed behavior is unconscious. These behaviors are inherited, as opposed to learned: they are "burned… so deeply into the circuitry [of the brain] that we can no longer access them."[11] Consider, for example, our innate fear of snakes or spiders, or how we defer to the dominant group member. The list may extend to whom we find sexually attractive, our fear of the dark, the capacity for empathy, and the propensity to argue, become jealous, seek solutions, and avoid incest – all of which are automatic, fast, and effortless.[12] Another influential "instinct" is the **binary instinct**. The binary instinct reflects our innate

tendency to create order by oversimplifying the world, dividing it into two camps -- us and them.[13] When we create an "other," we categorize a group as different and apart from ourselves, giving ourselves license to demonize and dehumanize its members. In other words, speed and efficiency often come at a price – they can hurt.

- *Implicit Re-actions*. **Implicit memory** has been described as something your brain has knowledge of but your mind can't access.[14] It holds many of our established **mind patterns**, and is a source of re-activity. Examples of how the implicit past can influence the present without our knowledge include the following:
  (i) The brain constructs mental models in the form of "if/then" rules that we apply every day[15]: if he nods, then he agrees; if she doesn't make eye contact, then she's guilty, etc.
  (ii) **Defenses** can range from ignoring a situation to projecting our faults onto others. While some are conscious, the goal is the same: to "build a fence around our awareness" so we don't have to face what scares us.[16]
  (iii) **Heuristics** are mental short cuts or rules of thumb we unknowingly apply to "reduce the effort associated with a task."[17] We apply a **representative heuristic** every time we notice a similarity between two objects and infer the first acts like the second[18]; and an **availability heuristic** when we judge something based on how easily examples of it come to mind.
  (iv) For those we are close to, we carry a historical portfolio of implicit assumptions about how they are and what sets them off, often preventing us from seeing a change in them.
- *Stereotypes*. We all **stereotype**; it's an economical way of "organizing and simplifying" a complex social world.[19] Stereotypes are a form of pattern, and at some point they helped our ancestors quickly distinguish friend from foe. Every day we "unconsciously, automatically, and immediately" categorize people along

- such lines as race, age, and sex.[20] Stereotypes are not always negative, but they become a problem whenever we adopt them blindly; ignore individual differences; use them to justify an initial belief; and/or apply them to make other people "less than."[21] Once formed, it takes real effort to work through our prejudices. Research has shown that when we are prejudiced and try to appear *not* to be, the cognitive burden literally decreases intelligent thought.[22]
- *Priming*. **Priming** originally referred to how the processing of a given stimuli readies us to respond to the same stimuli at a later point.[23] Today it generally refers to how cues too subtle for us to notice affect our judgments, thoughts, actions, etc. For example, in one study, students primed with words "consistent with the stereotype of old people" were found to walk "significantly more slowly" as they left the test site.[24] Moods also seem to prime our judgments: when we are in a positive mood, our judgments tend to be more lenient; in a bad mood, they tend to be more negative.
- *Biases*. The sheer number and diversity of human biases is startling.[25] Biases can be unintentional and are essentially undetectable to us.[26] The **fundamental attribution error** refers to the tendency to blame the behavior of others on their character, while attributing our own behavior to situational or external factors; the **egocentric bias** is the tendency to assume that the view we hold of the world is the one true view, and presume others share this view; and, the **belief perseverance bias** is the tendency to focus on facts that support our existing beliefs and ignore those that don't.[27] In addition, we prefer inaction to action (**omission bias**); tend repeat a first choice over a series of decisions even when our preferences change (**status quo bias**)[28]; and often reconstruct the past to fit into our present knowledge (**hindsight bias**).[29] Even experts have been found to "rapidly categorize...a novel situation" as something familiar to them, and proceed accordingly.[30]

Interestingly, one of the "most pervasive biases" is that we all assume we're less biased than the average person.[31]

THE NEW UNCONSCIOUS

By now it should be easier to appreciate how most of our cognitive function happens outside conscious awareness. More recent research, however, suggests that the percentage isn't 95% -- it's 98%.[32] In the past several decades, the list of "social phenomena...that are wholly or in part a function of automatic processes continues to grow," and many have been found to be "governed by a combination of automatic *and* controlled processes" (emphasis added).[33] It has also been found that "cognitive processes are not dependent on consciousness...[in fact] consciousness depends on *unconscious* cognitive processes.[34] What's being called **the new unconscious** covers the more recently identified behaviors that can occur unconsciously:

Figure I
The New Unconscious

Recently Identified Behaviors Triggered or Enacted Unconsciously	**Studies have shown:**
Complex Social Behavior	• Our capacity to read the mental states of others and make certain assessments about social interactions may derive from "an unconscious conceptual framework" that draws on past experience[35] • People are influenced by others' expectations even if they are not aware of them[36]

	- We automatically behave as we perceive -- we laugh when other people find something funny, change our behavior to fit into a group, and mirror others to get them to like us[37] - We unconsciously mimic the posture, gestures, etc. of another person, especially people we care about and those who may be of potential benefit to us. We do not tend to mimic our adversaries[38]
Self-Control	- The fact that the prefrontal cortex and the neocerebellum are connected suggests that the unconscious may be involved in "automated programs of executive control"[39] - People can unconsciously "monitor and correct for bias" in their judgments, just as they can do so consciously[40]
Imagination	- Imagining conditions that might have been, such as "what if" scenarios, can be either automatic or intentional[41]

Problem Solving	• "Efficient problem solving, performance enhancement, and affective coping may develop from intentional, deliberative thought" that is repeated and becomes automatic[42] • Discerning similarities, making comparisons, and drawing certain inferences can occur spontaneously in cognitive processes[43]
Insights	• The process that yields an insight – as opposed to the awareness of the insight itself – does not require conscious awareness[44]
Judgments Based on Glimpses	• From brief social encounters, people can be very accurate in their assessment of "legible" traits (those that are easy to observe), such as agreeableness[45]

HABITUAL BEHAVIOR

Habits are often unconscious too – it's estimated that more than 40% of the actions that people perform everyday are habits.[46] Habits and habitual behavior are important to helping professionals because they dominate everyone's behavior to some extent, and for some, they can be destructive. Habits are usually triggered by a familiar setting. They generally include a *pattern* or series of steps assembled into a routine that becomes automatic; a *cue* which serves as a trigger for the automatic behavioral routine; and a *reward*, especially one that grows so intertwined with the cue, it creates both desire and craving.[47]

Habits are a form of conditioning; they are learned associations. The greater the reward, the more incentive there is to recall the routine, and the more habitual the routine becomes:

- Habits can be described as *decision-making without thought*: biases, stereotypes, heuristics, body movements, states of mind, etc. can all be habitual;
- Like other unconscious influences, many habits begin as conscious, and become unconscious through repetition;
- In a habit, situational cues can be as important as the reward: an estimated 45% of what we do every day is done "in the same environment and is repeated."[48] Ultimately the routine gains the power to override intentions;
- Habits are often seen as a character flaw, but patterns unfold automatically: cues trigger cravings that cause routines which result in a reward;
- To kick a habit requires effort. For willpower to be effective, one needs to be consciously aware of the habit and have a plan for overcoming it; and,
- Stress, cognitive load, and overuse all undermine willpower. As every dieter knows, willpower is at its peak in the morning and dissipates throughout the day and through use.[49]

Not all habits are bad. Habits routinize what would otherwise be so labor-intensive we'd still be in our morning shower trying to figure out how to pick up the soap. But *unwanted* habits cover a wide range: according to one study, nearly 23 million Americans – almost one in ten – are addicted to alcohol or other drugs.[50] Even checking your email can give the body an endorphin-charged reward. The **basal ganglia** are key players in habitual behavior. The basal ganglia are able to:

- Tune into information about internal states, goals, and purposes;
- Recognize similarities between situations and across circumstances, and generalize from them;

- Identify "reward characteristics" and facilitate "reward-driven association learning"; and,
- Focus attention, and select and pursue a particular course of action.[51]

The basal ganglia work with the **orbitofrontal networks** which motivate and guide behavior, and the **anterior cingulate cortex** which monitors success or failure in obtaining a reward.[52] The circuitry involved in forming and maintaining a habit also includes the synthesis of **dopamine**. The release of dopamine sends the message "Do it again," creating a craving for the reward. Dopamine neurons learn to gauge reward levels: whether the reward is as expected, not forthcoming, or less than or better than expected. Dopamine release has been found to occur under both stressful *and* rewarding situations, suggesting that dopamine has more to do with **prediction-error** than pleasure: it compares anticipated rewards with actual ones and directs future activity towards the most rewarding pursuits.[53]

There was a time when a simple shift of goals or intentions was thought to be sufficient to change a habit. This is no longer the case. Many books on the market today focus on willpower: understanding it, applying it, and learning tricks for getting the prefrontal cortex to do "the harder thing."[54] One article suggests making self-control an *unconscious* habit.[55] Still other books say the best way to change a habit is to consciously identify it, along with the cues and rewards that trigger the habit, and replace it with a healthier habit; modifying the setting in which it occurs and ensuring the new habit provides an equivalent amount of reward or relief.[56]

TAKE A *CONSCIOUS* BREATH

In addition to these unconscious influences, the **wisdom** of the brain suggests we each carry **a set of unconscious assumptions about living** that are rarely mentioned and almost never examined. We each assume that life should be good; that we should have the good without the bad, endure a minimum amount of pain, and little or no suffering. We assume that conditions are inherently dualistic, problems can be fixed once and for all, and there's a way things "should be." We assume we should get what we want; that good things should last...and we will last. Whoever said we were entitled to a "good life"? Better yet, who defined it?

If a full 98% of cognitive function is unconscious and a good percentage of that is not even accessible to consciousness, chances are we barely know ourselves. What's more, what we *do* know is inferred from observing ourselves or told to us by others; filtered through our unconscious layers or the unconscious layers of others. So why don't we doubt ourselves more? Because the brain avoids uncertainty and finds doubt uncomfortable. Besides, we're probably not consciously aware that we should.

Recommended Reading
Subliminal, by Leonard Mlodinow and Incognito, by David Eagleman
The Power of Habit, by Charles Duhigg and The Willpower Instinct, by Kelly McGonigal

A
Closing Thought

A Neuroscience Question: How do unconscious perceptions, body movements, and instincts affect how we behave?

1. They unknowingly influence our actions[57]:
   - People rate food that is "lyrically described" -- such as velvety mashed potatoes -- as better tasting than the same food described more generically;
   - The size of a container affects how much we eat: in one study, by doubling the container size, participants consumed 30-45% more of a snack food; and,
   - Studies have also shown that sunshine affects how we behave, attractive packaging influences the products we buy, and the same wine tastes better if we think it's more expensive.
2. They unknowingly influence our re-actions:
   - A hunch can be the difference between what we consciously perceive and what the unconscious takes in [58]...a "funny feeling" about someone or sense that "something's wrong";
   - We cannot be introduced to another human being without feeling "some immediate attraction or repulsion."[59] The brain decides what we think before we're consciously aware of it; and,
   - We react more strongly to something that happens as a result of a commission than an omission,[60] and favor people who have traits similar to our own.[61]
3. They unknowingly influence our cognitive and emotional behavior:
   - We generally feel more empathy and show greater responsibility for victims we feel close to, including those who share our beliefs and attitudes[62];

- We treat attractive people differently. Even infants "prefer" faces considered attractive by adults, irrespective of race, age, or gender[63]; and,
- We spontaneously compare new situations with the past. This kind of "quick, cheap, similarity processing" is always active and influences how we represent and reason about a new situation.[64]

4. They unknowingly impact our social relations:
   - People readily give off signals: we pick up on their mood, disposition, social status, and nonverbal cues. For example, we subliminally "adjust our gazing behavior to match our place" in the social hierarchy and do so "with numerical precision"[65];
   - We spontaneously form impressions about people, often with minimal information. In the first tenth of a second, we can make a snap judgment about a person's competence, aggressiveness, even likability.[66] A 33-millisecond exposure to a face is enough to assess whether a person is trustworthy.[67] According to more recent research, the brain "judges the trustworthiness of a face before we even consciously see the face"[68]; and,
   - We unconsciously discern from certain "tie signs" whether two people are a couple, when it's time to pause in a conversation, and how to switch between the roles of speaker and listener.[69]

21

Making a Conscious Effort

Consciousness is difficult to describe. It typically has something to do with awareness, often the awareness of being aware. It tends to be a desired state, which tells us we are not always in a conscious state; it must be both *sought after* and *engaged*.

Consciousness is often discussed in terms of intelligence, free will, and/or self-control, each of which we will examine. It may also be tied to knowing, as in "to have a knowing." There are no "consciousness neurons," no region of the brain that acts as the "seat of consciousness," nor does consciousness sit atop the neural hierarchy.[1] **Re-current processing**, which involves the **prefrontal cortex**, has been found to be necessary but not sufficient for consciousness.[2] In addition to the prefrontal cortex, areas of the **parietal cortex** and the **anterior cingulate cortex** are "more associated with consciousness"[3]; the **thalamus** and parts of the **medial temporal lobe** are involved as well.[4] All told, it's likely the brain creates consciousness as it does any other complex mental function: it works together as a whole and consciousness **emerges**. Whereas unconscious mechanisms drive most of lives, consciousness allows us to notice, attend, magnify, focus, invent, and create.

Despite the vagueness surrounding the concept, everyone seems to have an opinion on what consciousness is:

- Consciousness is attention. It is a precursor of the mind, uniquely human, and responsible for human innovation.[5] It integrates and

summarizes information, tips the scale in decisions, amplifies what is relevant, and allows judgments to be made with confidence;
- It's a feeling of aliveness, the awareness of what is in working memory,[6] necessary for complex problem solving, biased by attention, and part of noticing and change;
- Consciousness is, at some level, independent of the brain -- the **mind** is a state of consciousness. It is subjective experience, the ability to look inward, and our capacity for knowing;
- It assembles and oversees the unconscious parts of ourselves[7]; its purpose is to oversee behavior. It maintains a "coherent story, a self-concept"[8] and enables deliberative thought;
- It is our shared experience. The mind is more than consciousness because mental processes, like shame for example, can exist outside of it.[9]

Consciousness has also been described as our unified experience, a given, what it means to have a mind; as Pure Intelligence.

CONSCIOUSNESS AND INTELLIGENCE

Consciousness has long been associated with human intelligence and is sometimes regarded as setting humans apart from other animals. It allows us to be "explicitly" aware of our personal experience, which has been said to be the "ultimate manifestation of intelligence."[10] There are many different kinds of intelligence including linguistic, logical-mathematical, spatial, inter- and intra- personal, kinesthetic, musical, and naturalistic.[11] **General intelligence** is associated with the amount of gray matter in particular areas of the frontal lobes[12] and with the size of working memory. **Fluid intelligence** refers to our capacity to analyze relationships, make inferences, and apply knowledge to solve new situations.[13] **Crystallized intelligence** is the accumulation of skill- or strategy-based knowledge, while **metacognition** involves being aware of the effects of our actions and the ability to work with abstract concepts.[14] The link between consciousness, intelligence, and *language* continues to be debated in the neuroscience community.[15]

Intelligence is sometimes related to knowing. Knowing is a neutral faculty; its effects depend on how it's used. We can know something is unjust or come to know our own deeply-seated **mind patterns** by bringing them into awareness. Alternatively, we can know precisely what someone else needs to do or "teach and speak as a way of staking out our position."[16] At issue is whether our knowing encourages or infringes upon another's freedom to think; also at issue is our comfort level with the unknown.[17] The abundantly connected **prefrontal cortex** is necessary for performing most of the skills related to intelligence.[18] As such, when we're under a cognitive load, it's harder to be consciously aware, to control ourselves, or to think clearly. Just as consciousness does not have ultimate veto power over our behavior,[19] intelligence is no guarantee of a good decision. Moreover, we can be both conscious and intelligent, and still make bad choices.

Attention is often linked to intelligence and consciousness. Attention is the sum of "two antagonistic and complementary processes": one is inclusive and focuses, while the other is exclusive and inhibits distractions that may draw our attention away.[20] Once something moves outside conscious attention, it is virtually invisible to us. The prefrontal cortex makes selective attention possible; working memory then processes what we train our attention on. **Creativity** involves bringing together "disparate ideas in new and useful combinations."[21] While creativity is often mentioned alongside consciousness and intelligence, it is unique in many respects:

- Creativity works best *without work* – with a mind that is at ease and free from distractions;
- It requires just enough attention, while still allowing unrelated ideas to surface and coalesce. Also, too much prefrontal control can inhibit imagination, emphasize conventional solutions, or make us fear criticism[22];

- Both highly intelligent people and people with low levels of intelligence can be highly creative.[23] Creativity requires the unique combination of drive and discipline, *and* mental freedom and playfulness;
- Creative ideas produced by individuals are of greater quality than those produced by groups[24]; and,
- Creativity is a slow, meandering process, the *outcome* of which may suddenly come into conscious awareness.

Insight is closely related to creativity. An insight "solves a problem or reinterprets a situation in a non-obvious way."[25] It occurs best when people relax and focus their attention inward.[26] In studying **the neural basis of insight**, researchers have found that that about one-third of a second before the moment of conscious insight, there is a "distinctive flash" of **high frequency gamma waves** from the right anterior **superior-temporal gyrus**, where associations are made and solutions are assembled; almost always preceded by a change in **slower alpha-band** activity over the right **occipital** (visual) **cortex**.[27] This may mean that "when a weakly activated problem solution is present in the right temporal lobe, a temporary reduction in interfering visual inputs facilitates" its retrieval.[28]

CONSCIOUSNESS AND FREE WILL

Consciousness has to be engaged. It is often tied to our sense of **agency** and an independent self, and seen as the cornerstone of free will. Consciousness can be described as a state of *self*-awareness. Situated by episodic memory in a specific place and time, and accompanied by an evolving narrative, the self becomes *the object* of our awareness. Our subjective experience is one of a self that is at the center of what feels like a unique and private life. The self sees the world as a unified whole and is not aware that experience is filtered and indirect. It also has the power of *intentionality*.[29] According to the **theory of apparent mental causation**, "we experience ourselves as agents who cause our actions when our minds provide us with previews of the actions that turn out to be accurate..."[30] This is especially true of actions "caused by *controlled* processes"[31] (emphasis added) which allow us to draw inferences about how this occurs. We thereby "accumulate a picture of a virtual agent," one that freely and consciously wills our actions.[32]

But when free will is put to the scientific test, conscious awareness is not in the driver's seat. In Benjamin Libet's experiment in the 1980's, volunteers were asked to "spontaneously" lift a finger and state precisely *when* they had the urge to do so. Researchers found that brain activity related to preparing to move began a "good third of a second, on average, before the subjects consciously believed they had decided" to do so.[33] In a more recent study, "The decision that participants were going to make could be predicted...[a full] 7 seconds before they were consciously aware of that decision."[34] Add to this the belief that 90% of the input into the cerebral cortex comes from within -- from "internal neural processing"[35] -- and rather than free will, the unconscious filter of past is driving our experience and behavioral output.

This is *not* to say that we do not initiate and execute our own actions. Rather, at least *some* of our decisions are "prepared preconsciously" before we're aware we've made the decision.[36] As we've seen, a lot happens in the brain outside consciousness awareness and those processes shape behavior. And we can't forget that the perception of the mind and the self as a virtual agent is *a construction* – a "deeply important" one to be sure, because it allows us to function as a person in a social world. But it's an "*experience* of agency," not "the direction perception of an agent."[37] In sum, actions "may depend less on conscious intention than we think" because "by the time consciousness kicks in, most of the work has already been done."[38]

David Eagleman, in his book Incognito, discusses the implications of this on criminal responsibility. He suggests that "it no longer makes sense" to ask whether "a person's biology" or the person is accountable for his/her actions "because they are one in the same."[39] Culpability, he argues, is the *wrong question*: by virtue of his/her actions, the criminal has a "brain abnormality" and "should be treated as incapable of acting otherwise."[40] According to Eagleman, rather than blame, we should be asking, "What the hell do we do with this guy?" Other authors counter that the reason we are ultimately accountable for our actions is because following unconscious processing, there is a brief conscious moment of choice. Alternatively, the **wisdom** of the brain suggests we are **accountable because we <u>can</u> bring the results of unconscious processing into conscious awareness**. We usually don't, but we can. It's part of what makes us fully human.

CONSCIOUSNESS AND CONTROL

Consciousness is often equated with control over the events in our lives and over our self. It is, as a consequence, closely aligned with the **prefrontal cortex**. The prefrontal cortex (PFC) is responsible for planning, higher level decision-making, analysis and abstract thought, inhibitory control, and goal-directed behavior; it plays a key role in social interactions, imagining the future, comparing multiple contingencies over time, language, dealing with uncertain situations, and devising new takes on old strategies.[41] With respect to consciousness, the prefrontal cortex participates in **recurrent** (also known as **reentrant**) processing,[42] where the brain feeds information back from higher to lower areas of the brain, continuously updating itself as it goes along. Recurrent processing is thought to "accompany conscious experience,"[43] and may be part of the **neural correlates of consciousness**.

The prefrontal cortex makes selective attention possible; it supports **working memory** which keeps our attention trained on what's important; and it is heavily involved in self-control. Self-control, or **self-regulation**, includes inhibiting impulses, controlling emotions and managing stress, delaying gratification, resisting temptation, choosing the appropriate behavior for the setting, and other difficult tasks. But as we've learned, the prefrontal cortex is not an executive controller; its capabilities depend on its interconnectedness, especially with subcortical regions of the brain. What's more, top-down control isn't always a good thing: Dr. Daniel Siegel talks about the "potential tyranny of top-down domination," the "flow of summarized life memory from our busy, knowledgeable, and future and past-minded...cortices" which can "obliterate, or at least obscure, the freshness" of a more bottom-up or visceral experience.[44] Regardless, top-down control takes conscious effort; it has to be deliberately engaged, monitored, and repeatedly exerted. As such, it's not entirely reliable.

Sometimes when inhibition isn't enough, people want to **change**. Consciousness, control, and change are sometimes linked together. **Second-order change** is especially difficult because it requires changing *our nature*: to *permanently* slow our breathing, modify our "natural" activity level or goal orientation, alter an emotional or physiological setpoint, change our self-talk or a relationship with another, etc. ...and then progress to the next second-order level change. **Mindfulness** is one way to achieve second-order change *and* grow in conscious awareness. Mindfulness trains the mind to "pay attention" to the present moment in a nonjudgmental way.[45] It helps quiet the mind, fosters greater objectivity and emotional stability, decreases reactivity, and can increase both self- and other-awareness. Research has found that long-term practitioners of certain kinds of mindfulness meditation have literally changed how their brains work.[46] Mindfulness meditation, however, is not for everyone. It is definitely not the path of least effort. In addition, some types of practice are more effective than others. Mindfulness does not ensure happiness: after mindfulness practices, people can feel happy or sleepy, thankful, refreshed, prayerful, disengaged, etc.[47] And like any new habit, mindfulness starts with a desire; at least initially, it requires consciously overriding old mind patterns and strengthening the circuitry necessary to support a new habit. More research is needed to fully understand the benefits, duration, and applicability of such practices.

CONSCIOUSNESS EXPLAINED BETTER[48]
While consciousness has many meanings, several ideas might help better explain the concept:

1. Consciousness can be seen as adaptive: evolutionarily, it probably emerged slowly and it is likely to be present in humans and other animals[49];
2. The reason for consciousness may be that it "facilitates reflection," through which we can learn and change[50];
3. Consciousness is derived from the physical brain, although is not localized in the brain and depends on widely distributed networks;
4. Throughout the day, we probably switch between consciousness and unconsciousness, spending much more time in the latter. Even our conscious moments are shaped by unconscious influences; and,
5. Consciousness is neither good nor bad. It can lead to deeper levels of self-awareness or be "engaged ruthlessly and innovatively to achieve our irrational unconscious goals."[51]

Consciousness can also be understood outside the traditional Western paradigm...where it is infused into the mind at birth; where the brain is a physical instrument through which the mind works and the mind is the instrument for awareness. In this view, "intellect" uses reason based on the senses while intelligence is the "power of direct perception."[52] Here, the mind in always in motion and the goal is to *move beyond it*.[53] While human consciousness may be evolving towards a deeper awareness of our place in the universe, we remain with what one author calls a **primitive mind**. A primitive mind is prone to causally link phenomena (**associative thinking**); generalize from bits of information (**generalized thinking); categorical thinking** -- including the need to categorize and compare one's self; **self-centered** thinking; being fixated on the past or on one's immediate needs (**backward or instant thinking**); **selective memory**; and **state-specific reactions** where reactions depend on how one feels.[54]

Consciousness and unconsciousness cement our individuality; they form the **life of the mind**. The mind is where the self lives. The self has both conscious and unconscious parts around which the mind weaves the story of "me"; some of which we're aware of, most of which we're not.

And that's true for every one of us: just like our employees, clients, students, and patrons, we bring our primitive mind to work. This is why we can't be neutral; it's a biological impossibility. We can be neutral to the outcome and even neutral to the process, but we can't *be* neutrals. No mind is free from involuntary and unconscious influences.

Section VII

A Brain that's Also Social

22

Social Cognition

Human beings are social animals – we want and need to relate to others. To be social requires social cognition, it's how the brain and mind make sense of social interactions. Social cognition is subject to the same accumulating filter of past that guarantees the past shapes our present experience. It's also highly contextual: a person crying at a funeral will be understood differently than the same person crying at her daughter's awards ceremony. Social cognition is an **embodied** experience: we "read" others through their expressions, eye gaze, eye brow arch, mouth position, shoulders, posture and stance, hand gestures, vocalizations, felt feelings, touch, energy level, and resonance. The **wisdom** of the brain tells us that **to be in relationship means to develop the capacity to relate to everything**. It requires a brain with a distinctly social orientation, the continuous honing of our social cognitive skills, and the capacity for self-reflection -- a faculty still evolving in the human species.

A DISTINCTLY SOCIAL ORIENTATION

Social cognition is the processing of information that guides our social behavior, interactions, and relationships with others.[1] It highlights the seventh basic principle of brain-building and operation: that *we are, and hence our brain is, increasingly social.* Like all complex brain activities, social cognition is built on **patterns**; in this case, on patterns of interaction within and among people. Through experience, we learn to automatically and reflexively read others; modify our behavior based on the reactions of those around us; imitate, simulate, and learn from other people; judge the nature, attitudes, and personality of those around us; and communicate and relate in accordance with cultural and social norms.

Social cognition involves the "perception of, attention to, memory for, and thinking about other people, and in a way that involves emotional and/or motivational processing."[2] As opposed to general knowledge or **cognitive intelligence**, **social intelligence** is more case-based than rule-based,[3] making it harder to learn and more subject to change. It is at the heart of motivation, communication, persuasion, conflict, cooperation, and empathy. The "simple" task of hearing another person's point of view requires that we first quiet our own opinions, then shift out of our ongoing first-person narrative, construct another person's viewpoint, and contemplate and compare it with our own. Social cognition is no simple task.

Social cognition is learned through relationships -- through the internalization of culture and the transmission of conventions, history, and narratives. It is an adaptive skill set:

1. Social cognition is a survival skill: those who can predict the behavior of others, blend into a community or culture, cooperate with fellow group members, speak the language, learn the mannerisms, meet the expectations of others, and mirror and resonate with a partner, go on to survive and reproduce.
2. Many of our social cognition skills are **unconscious**; pupil dilation, facial expression, posture, gaze following, touch, personal space, and proximity -- all work reflexively, outside conscious awareness. Moreover, we automatically gauge, attend to, and respond to others. It does not follow, however, that social cognition is always accurate.
3. Social cognition serves as both a predictive and corrective skill set. It is built on self-knowledge and self-perception, group interactions, and culture. There are two prerequisites for being a good group or cultural member: social cognition and self-control.

4. **Mirror neurons,** a kind of **motor neuron**, play an important role in social cognition. However the neural capacities we use to understand *actions* may be different from those we use to try to explain emotions and beliefs.
5. **Imitation** is a core skill in social cognition: we imitate others' behavior, **simulate** their mental and emotional states, learn skills through replication, and adopt a group mind, team allegiance, prejudice, and patriotism by imitating those around us.
6. **Theory of Mind** (ToM) is our awareness of our own and other's beliefs, desires, and intentions. It is a crucial part of social cognition. Theory of Mind may be an extension of our sense of agency, evolving from "preexisting neural networks" that served related purposes and had survival value[4];
7. Social cognition proves the brain is a **social organ**. Like any complex function, social processes rely on distributed neural substrates: subcortical regions of the brain are heavily integrated and one region can be involved in multiple functions just as one function can derive from multiple regions.

SOCIAL NEUROSCIENCE

Social neuroscience is the study of social cognition – it is the study of "all the ways in which human beings influence and are influenced by the presence, actual or imagined, of other humans."[5] It has particular interest in the neural functions that may serve or are correlated with social interactions.[6] Social neuroscience research has found that social cognition tasks "preferentially recruit a consistent set of neural regions" including the **medial prefrontal cortex** (MPFC), the **superior temporal sulcus** (STS), **medial parietal cortex (precuneus)**, and the **lateral parietal cortex**, especially the **temporoparietal junction** (TPJ).[7] As such, social neuroscience wrestles with "whether the processes that give rise to social cognition are a subset of more general cognitive processes, or whether specific social-cognitive processes exist"[8] – whether we have a brain that's also social, or a social brain.

Social neuroscience research has also found that:

- The neural systems that support **social** and **non-social thinking** are "quite distinct" and sometimes operate at odds with one another.[9] That means the social systems of the brain might be telling us one thing, while the non-social parts are directing us to so otherwise;
- Social and cognitive (non-social) intelligence may work as a "neural seesaw" – when one is engaged, the other becomes less activated.[10] So it's difficult to think socially and non-socially at the same time; and,
- There is overlap between the **default network** and the neural systems that support social cognition.[11] This makes sense because the default network is the system that focuses on self and others when we are not on task.

The areas of the brain that are especially involved in social cognition do not *cause* anything; they do not act alone, nor do they control us. They are highly interconnected, projecting to and receiving **projections** from many other parts of the brain, and have a distinctively social orientation. To begin to familiarize ourselves with these brain regions, we will first examine particular areas of the brain and how they relate to social activities:

Orbitofrontal Cortex and Medial Prefrontal Cortex (OFC and MPFC)

- Medial prefrontal cortex sub-serves all areas of social cognition
- Orbitofrontal cortex specializes in emotional processing, reward and inhibitory processes, real life decision making, self-awareness, strategic regulation, empathy, and appropriate social behavior.

Ventral Prefrontal Cortex (vPFC)

- Supports behavioral and emotional control/regulation

The Wisdom of the Brain

Ventral Prefrontal Cortex and Striatum

- Involved in reward processing

Dorsomedial Prefrontal Cortex (DMPFC)

- Thought to be involved in some mentalizing tasks

Right Ventral Lateral Prefrontal Cortex (rvLPFC)

- Supports taking a perspective other than one's own[12]

Septal Area

- Thought to help put empathy into action

Insula

- Reflects on emotional experiences
- Plays a role in assessing trustworthiness
- Participates in self/other distinction, identification of facial expressions in others
- Along with anterior cingulate cortex, activated with shame
- Along with putamen (part of basal ganglia), activated with disgust

Cingulate Cortex

- Helps translate intentions into action, willful control, overcoming habitual responses
- Plays a role in deception
- Part of basic circuitry for cooperation, and empathy

Amygdala

- Unconscious appraisal of others based on past experience
- Right side activated by losing, left by winning
- Helps discern emotional content of faces, fearful more so than angry faces

Periaqueductal Gray (PAG)

- Pain and attachment behavior

In addition, it's important to consider several social functions or tasks and some of the neural correlates that support them:

- *Being self-aware.* Supported by right parietal cortex (self in space), medial prefrontal cortex (MPFC), and ventral medial prefrontal cortex (vMPFC).
- *Recognizing one's self.* Supported by right prefrontal cortex (rPFC) and right parietal cortex.[13]
- *Deciding the applicability of self-descriptive information*. Supported by medial prefrontal cortex and precuneus (on midline of brain where the two hemispheres meet).[14]
- *Being able to self- reflect, having a sense of 'I', thinking about "who I am*. Supported by the medial prefrontal cortex.[15]
- *Theory of mind, mentalizing about the states of others*. Supported by medial prefrontal cortex, temporo-parietal junction (TPJ), precuneus/posterior cingulate, and temporal poles.
- *Mentalizing about others like me*. Supported by the ventral medial prefrontal cortex.
- *Mentalizing about others not like me.* Supported by the dorsal medial prefrontal cortex.
- *Imitating and mimicking others.* Supported by the mirror system -- mirror neurons are motor neurons that support action.

- *Displaying attention, having a self-concept*. Involves **spindle cells** -- spindle cells are known to be located in anterior insula and anterior cingulate cortex.
- *The notion of a self*. Generally distributed throughout the brain.
- *Processing faces*. Supported by lateral **fusiform area**, inferior occipital gyri, and posterior superior temporal sulcus (STS).[16]
- *Experiencing social pain.* Supported by dorsal anterior cingulate (dACC), insula, and right ventral prefrontal cortex.

A
Closing Thought

A Neuroscience Question: Is there a Social Brain?

Social neuroscience is faced with the question of whether we have a brain that's also social...or a social brain; or more precisely, "whether the processes that give rise to social cognition are a subset of more general cognitive processes, or whether specific social-cognitive processes exist."[17] If there *is* a social brain – if areas of the brain or neural substrates are uniquely and distinctly social – there's a tendency to view this as evidence of our innate cooperative nature and universal concern for others; that humans are on an evolutionary path towards world peace and social harmony... which we would all like to be the case.

But there is no social brain per se: there are no regions of the brain that are exclusively social; nor has a "tier" been added to the brain that serves exclusively social purposes. According to Dr. Matthew Lieberman, a leader in social neuroscience and author of the book Social, the neural systems that specialize in social cognition are relatively independent of those for non-social cognition.[18] Research has shown that in the case of *non-social* brain activity, there is usually a "significant increase" in brain region activity during a task, followed by "a return to baseline" when the task is complete.[19] Brain regions involved in *social* tasks however, often show "very little increase" above an "already elevated baseline." This suggests that we may be in a "perpetual state of readiness" when it comes to perceiving others as "social agents."[20]

One author sums it up this way, "We should be reluctant to look for neural structures that are 'for' social cognition. Rather we should consider social cognition as emerging from a complex interplay of many structures, in the context of development, of particular culture, and considering the brain as a system that generates behavior only through its interaction with the body and the social environment."[21]

23

Social Beings, Social Mind

As mammals, we develop in and through relationships. We are social by design *and* necessity. To be social requires a unique blend of mutual interest, trust, and ongoing self-regulation. As such, not only do we have a brain that's increasingly social...but an increasingly social **mind** as well.

THE ORIGINS OF HUMAN SOCIALITY

Human sociality begins with the basic building blocks of evolution.[1] The first is **individual selection**. The evolution of behavior depends on the survival of the individual, with the goal of passing on as many copies of one's genes as possible. The second building block is **kin selection**: because our kin carry our genes, we favor them as a way of passing on our own genes. The third is **reciprocal altruism**. By reciprocating with others, we -- as individuals -- succeed and prosper along with everyone else. Several conditions must prevail for reciprocal altruism to work: the group must be relatively stable; the species has to live long enough to give and receive back; and there has to be enough **social intelligence** for group members to recognize one another.[2]

The work of Robin Dunbar and others suggest that the human brain evolved "to handle the cognitive demands" posed by living in larger and increasingly more socially complex groups.[3] Dunbar found that the relative size of the **neocortex** correlates positively with the preferred group size of a given species. In the case of humans, we can manage a group of approximately 150 members; many human organizations tend to be around that size.[4] According to Dunbar, group members benefit from greater protection, shared tasks, and better, more diverse reproductive options. So long as the group remains relatively small, grooming holds it together, creating alliances among "grooming clique" members. Once the group becomes too big, language is needed to take the place of grooming in order to maintain social cohesion.[5]

Group living is thought to date back as far as 54 million years in our "family tree."[6] In fact, we share many of our social behaviors with our "primate cousins."[7] But social living is always a balancing act between individual wants and needs, and what the collective needs and expects. Membership provides protection and the capacity to undertake mutually advantageous cooperative ventures. We belong to collectives because they meet *our* needs as well as for the mutual benefits they afford. While it makes practical sense to be social, there is always a tension: between self and other, self-preservation and social belonging, control and cooperation, separation and connection. Some theorists suggest it goes deeper than that -- that culture, the self, and group membership all serve to frame, contain, and *inhabit* the human mind; to make us think, behave, and relate to one another in a particular way.

THE POWER OF CULTURE

Culture gives us ideas, art, and language. It gives us a sense of identity, a set of moral principles and rules, and a common sense of purpose. It defines appropriate behavior, teaches us what it means to be social, and structures social living. Culture helps order the world: it organizes our lives around schedules, routines, styles, mannerisms, standards, arrangements, and beliefs…**patterns** that create order through conformity. Culture creates an "us," and in doing so, a "them"; it is the **binary instinct** in action, reflecting our innate tendency to oversimplify the world by dividing it into two, often opposing groups. Conflict can be thought of as different ways of organizing reality. Today our differences are much more cultural than genetic: language, laws, institutions, and beliefs – from

God to destiny – separate us "in ways that our genes do not."[8] The "glacially slow pace of biological evolution"[9] has given way to the *inheritance* of culture: cultural innovation and transmission are now the dominant mechanisms of change.[10]

Culture sculpts the mind to a *particular culture*. It is the invention of ways to tie us together through common values and practices, of "creating a mind shared by many."[11] We learn to think, act, learn, interact, pray, believe, and behave according to the prescribed traditions of our particular culture. It's been said that as adults, we cannot function without our culture because it gives us what we need to live and relate to others.[12] One could say culture does a lot of our thinking for us.[13]

In the 1970's, thought-shaper Julian Jaynes proposed his theory of the origins of the "modern mind." He suggested that up until around 1200 BC, human beings had a "**bicameral mind**," a mind comprised of two separate chambers. It was a time prior to consciousness and a sense of self. Instead of a **left hemisphere interpreter** narrating our life story, people "heard voices that they took to be gods" which "directed their behavior."[14] According to Jaynes, these voices came from the right side of the brain, and had to be communicated to the language centers of the left side.[15] He saw them as the "germs" of both religion and civilization.

As internal and external pressures mounted, "the bicameral mentality gave way to what was simply a more efficient use of our brain – namely **consciousness**,"[16] and with that, human thought changed. People gained an inner life, a "psychological interiority"[17]; they acquired a sense of self, a world-narrative, and a "**representational system**…expressed [through] language and metaphor."[18] With consciousness, the right hemisphere orientation of the brain was "replaced by a new mentality," a more "**unicameral mind**" where a self occupies the mind.[19] This shift created the need for externally imposed and internally adopted means of control because the voices were no longer there to tell people what to do. It led to governments, religions, and culture, and individuals "equipped with selves" that served as the "socio-organization's inner voice by proxy."[20] All of this made culture stronger because its orders, directives, and rules became internalized *as mind*.[21]

More recently, author Mark Pagel, in his book <u>Wired for Culture</u>, made a strikingly similar point. He contends that culture is the best way of "making more people," and natural selection *selects for* culture-absorbing people.[22] He envisions culture as "something akin to a software 'operating system,'" that is "installed" by our parents and the society-at-large "without our consent"; where we have no choice but "to allow our culture to occupy our minds."[23] To these authors, culture is a trade: culture gives us protection and opportunity in return for *mind control*. It frames a "consensual understanding of the world,"[24] where members believe the world is this way and not that.

INDIVIDUALS EQUIPPED WITH SELVES

We are, of course, heavily influenced by and highly attuned to others. We learn our opinions from others; imitate their facial expressions and mannerisms. We mirror their likes and dislikes, preferences, biases, and attitudes. We acquire our **cognitive belief systems** from others and look to them to define what has meaning. Our sense of self is largely built by others and specifically, by the people who populate our lives. We are *not born* with a sense of self: over the course of "normal cognitive development," the human mind constructs it.[25] It's entirely learned.

The story of "me" provides the blueprint which is groomed and grown in a particular culture into a full-fledged sense of self; "an internalized conceptualization of agency, imported by socialization processes."[26] The self is a series of thoughts constructed from the stories we are told that we adopt as our own. It reflects the brain's conservative nature. The self is not real -- as we've said, it has the "same status as a belief or theory."[27] It is a set of fixed ideas, a reality we devise, a large part of which is in our head. The self helps modulate our fears and believes it is more central in the world than we actually are. Because it is not real, the self is vulnerable on almost every front: it is easily threatened, defensive, re-active, self-focused, and other-attuned. Most importantly, the self is defined by where we fit – by our family, the groups of which we are a part, our culture, nation, church, and school -- by where we *belong*.

Dr. Matthew Lieberman, a social neuroscientist and author, suggests that while we like to believe the self makes us special, it's actually "a cleverly disguised deception" that allows the beliefs and values of a particular culture to be "internalized and treated as personal" without our knowing this has occurred[28]; it's how "society becomes subjectivity."[29] Lieberman sees this as a good thing: by encouraging us to share the same beliefs as those around us, we are brought into greater harmony and able to get along better with others. Certainly from an evolutionary standpoint, equipping individuals with selves makes sense: it's far easier to navigate the world with an "I" to organize life around and live through. On the downside, as a cultural construct, the self can serve as a "personal tool kit of command and control."[30] When the mind is inhabited by such a self, it fuels competition among individuals and the groups we belong to...

GROUP MEMBERSHIP

Groups define who is inside and outside the cultural circle. Human beings follow a general pattern: we tend to keep to ourselves and break off into distinct societies. In fact, we've had an "innate tendency throughout history to divide and form into new groups, distinct from others, and just as soon as the environment will allow it."[31] We do so because of our innate need to belong. The need to belong makes us willingly cooperate with fellow group members and obey the rules because we have come to believe – through culture, others, or perhaps even natural selection -- that the good of the group is in our best interests as well.[32]

Groups provide us with a feeling of certainty and permanence. They afford us protection in numbers and the capacity to do what no single individual could ever accomplish. Group membership helps shape our identity; it defines and refines how we see ourselves. The desire to be a member is so strong that studies have shown we "categorize ourselves as group members on the flimsiest of pretexts."[33] We like to see ourselves as the same as other members of our group, and non-members as outsiders.

Group membership gives us belonging in exchange for conformity. To ensure conformity, evolution has made sure that "the mere possibility of being judged" by others significantly increases our tendency to abide by society's rules and values.[34] In one study, the mere *picture* of a pair of watchful eyes posted in a work break-room increased *by 276%* the amount people paid into an honesty box.[35] Groups have a full battery of weapons to ensure members play by the rules: rewards, ridicule and mockery, bullying and public shaming, anger, conflict, ostracizing, sanctions, expulsion, and war.

When we form into groups, we behave in particular ways, as individuals and collectively: we have more empathy and assume greater responsibility for fellow group members; we categorize others and base our expectations on associated stereotypes; we gravitate towards people who share our opinions and prefer people who praise us[36]; we process information so it fits with our preconceived notions, and willingly cooperate when our interests are also served. To be certain, human beings are social animals. Groups persist because they help perpetuate culture and because individuals get something out of them: we get our needs met.

SELF-REGULATION

Underlying culture, our sense of self, and the groups we belong to, is the need for **self-regulation**. Dr. Lieberman calls it the price of admission. **Inhibition** is a critical part of self-regulation. It refers to the process by which "people initiate, adjust, interrupt, stop, or otherwise change thoughts, feelings, or actions," in order to realize personal goals or maintain "current standards."[37] To stay in the group's good graces, members need to "inhibit their impulses, stifle their desires, resist temptations, undertake difficult or unpleasant activities, banish unwanted and intrusive thoughts, and control their emotional displays."[38]

When we exert self-control, we are trying to match our behavior to the expectations of the group. Lieberman, among others, suggest that society may be the "primary beneficiary" of this because it reinforces group identity and promotes cohesion.[39] In other words, self-regulation sustains the order that in turn, preserves the culture. The point is this: as the brain has grown increasingly social, we have grown increasingly controlled and controlling. This creates a tension that manifests in the mind. As social beings, we accept control for the order it affords as well as the protection and opportunity we hope it provides. It fosters a feeling of certainty, but *it's only a feeling.*

Evolution has yet to invest heavily in the **self-reflection** necessary to see through our differences and distinctions. But the **wisdom** of the brain tells us **we are not our mind** – because *it* can be observed. Rather, what matters lies beyond culture, the self, and group membership, in the power of *individuation* – in seeing others as the unique individuals they are…sitting among other, equally unique individuals.

A
Closing Thought

A Neuroscience Question: What is the truth about penguins?

It's often said that because we are social beings, humans need to belong; and if we *need* to belong, it follows that we are naturally cooperative because cooperation secures membership. Further, that at the heart of both cooperation and belonging is human altruism -- not just **reciprocal altruism** but an "unbounded" altruism implying that humans are by nature, good. Looking at this more closely:

1. Human beings are deeply connected to other humans, "and most actions are reciprocal and in the interest of both parties"[40];
2. One well-known study found that dopamine increased when both parties cooperated...but all the participants in the study were female[41];
3. Studies have found that when participants observed unfair players in pain, there was decreased activation in brain areas involved in the emotional aspects of pain...but only in male participants[42];
4. In competitive situations, observing other people's joy has been found to increase distress, while observing their pain results in positive emotions...in both sexes[43];
5. We exert greater self-control when we imagine what people who are important to us would think of us.[44] We tend not to mimic our adversaries or feel their pain[45];
6. Providing social support feels good. Specifically, it feels good to help those we care about[46]; and,
7. **Oxytocin** has been found to increase bonding towards in-group members, and increase hostility towards out-group members.[47]

The 2005 documentary <u>March of the Penguins</u> showed penguins taking turns, each spending time near the center of the group where it was the warmest. The narrative stated that at least at this time of year, they functioned as a "united and cooperative team…merging their bodies into a single mass."[48] The problem is *this isn't true*. Penguins don't take turns, nor do they willingly leave the warm center. Each penguin tries to push and shove his way to the center and the strongest, most aggressive penguins win.

Humans, too, cooperate *and* are selfish; we give and we receive. We are afraid of and shaped by others; built by and constricted by our relationships; programmed by and protected by the collective; controlled and controlling. And while belonging may be more important to some people, when push comes to shove, evolution says "I fight for me."

There is definitely an evolutionary value to belonging. But we must ask, what do we mean by *belonging*? Does it mean to fit in, to mold ourselves to a set of preconfigured expectations? Does it mean we have a stake in something; that we identify with a particular group or purpose? Does it mean we connect with others, and if so, is there an expectation that we will be like them; that we will somehow change in order to secure that connection? Brene Brown, who wrote Daring Greatly, defines belonging as requiring us to be who we are; as being loved or accepted or connected with *as is*.[49] Psychologist Adam Waytz put it this way: "Being social is far from easy, automatic, or infinite…our altruism is not unbounded, it is parochial."[50] We are driven by both social- and self-interest, and as with all things, we have to take the good in ourselves with the bad.

24

Reading Other People

When we use social cognition, we do so in conformance with cultural norms, consistent with our sense of self, and in keeping with group expectations. Social cognition is a form of knowing. On the plus side, despite the fact that we can only know another person indirectly and we barely know ourselves, we continue to try: social cognition is the sum of our efforts to know our self and others. It is at the heart of every helping profession. On the minus side, it's a skill set that requires constant attention. The **wisdom** of the brain is to **know enough to assume we might be wrong**; to be comfortable knowing our interpretation of a fellow human being may well be incorrect and requires their input.

PROCESSING FACES

Faces probably give us more immediate information about another person than any other feature. Babies are born with a preference for human faces; they fixate on their mother's face and learn to link gazes as part of the parent-child bonding process. As we grow, we learn how to read faces, connect with and generally relate to others. Research has found that basic facial expressions do not vary culture to culture, so much so that the facial expressions for the happiness, anger, fear, disgust, sadness, and surprise are generally considered universal. Moreover, "people can recognize emotions in faces from all cultures at striking levels of accuracy."[1] Even children who are blind and deaf, and know nothing about communicating emotions, "create spontaneous facial expressions that greatly resemble" those of other children.[2]

We use faces to recognize people, process their emotions, predict their intentions, assess dominance, perceive a threat or danger, and detect group membership. *Seeing* faces activates the **occipital lobe, temporal lobe, occipital-temporal junction**, and the **inferior parietal lobe**.[3] *Processing* faces involves the **fusiform face area** (also known as the lateral fusiform gyrus), the **inferior occipital gyri**, and the **posterior superior temporal sulcus**, and other regions that respond to emotional and social information such as the **amygdala** and the **precuneus**.[4] When we *recognize* a familiar face, there is heightened activity in the fusiform face area as well as regions linked to mentalizing, episodic memory, and emotional responses.[5] Interestingly, the fusiform face area does not process upside down faces – the brain treats them as objects.

The two **hemispheres** of the brain together produce and assess facial expressions. The right hemisphere is especially adept at detecting upright faces and is better than the left at recognizing unfamiliar faces.[6] While *spontaneous* facial expressions can be triggered by either hemisphere, only the left hemisphere can produce *voluntary* facial expressions; people can tell the difference between a spontaneous smile and one that's produced voluntarily. The **amygdala** plays an important role in detecting facial expressions, both fearful and positive ones; it is, however, more attuned to fearful than angry, disgusted, or surprised faces.[7] We also rely on the amygdala to assess another person's trustworthiness. In terms of happy faces, research has found that people who are more extroverted process happy faces as a reward, whereas introverts feel less of a reward when approaching others.[8]

PROCESSING SOCIAL CATEGORIES

We can "take in" another person in a matter of milliseconds. We "automatically, unconsciously and immediately" categorize people along such lines as race, age, and gender.[9] Within a tenth of a second, we can make a snap judgment about a person's competence, aggressiveness, even likability.[10] No-one is immune to being categorized or categorizing others:

- We automatically pick out similarities among people based on visible features, and use multiple cues to categorize others including skin color, weight, height, age, youthfulness, political affiliation, country of origin, mental competency, etc.;
- Information about race and gender is considered very early in the categorization process.[11] Studies have shown that we process "race effects" in as little as 120 milliseconds and gender in 180 milliseconds, after the stimulus is presented[12];
- When we think about people we perceive as similar to us, we use our own personal knowledge to understand them. Those dissimilar to us are harder to read, and we must rely on the more dorsal area of the medial prefrontal cortex[13];
- When we categorize people based on **implicit prejudice**, it engages the neural systems related to threat detection and response. When we use **implicit stereotypes**, the process is more conceptual and hence the systems are more cortical in origin[14];
- The drive to control our biases can come from within or from those around us. Research has found that the faster we categorize "outgroup faces," the less we "individuate" them[15] -- the less we see them as a person with an inner life and a mind...just like me.

PROCESSING SOCIAL PAIN

According to the work of Naomi Eisenberger and her colleagues, the brain processes social pain in very much the same way it does physical pain.[16] Specifically, the brain treats social rejection like a painful event. The key player in this is the **anterior cingulate cortex** (ACC) which monitors for conflict in cognitive and emotional matters, and warns the prefrontal region when there's a problem that requires attention. Evidence has shown that the dorsal part of the ACC (dACC) is involved in the "distressing components" of both *physical* and *social* pain,[17] leading researchers to suggest that the social attachment system was built "onto or developed out of" the physical pain system.[18] The **right ventral prefrontal cortex** (rVPFC), which is part of the neural systems involved in physical pain, has also been found to show greater activity in social pain. It's thought that the dorsal anterior cingulate cortex helps produce the subjective experience of pain, while the right ventral prefrontal cortex helps "down-regulate" the experience.[19] The finding that there is some neural overlap between the systems that support physical and social pain demonstrates that the brain takes social threats just as seriously as it does physical ones. It also helps explain why social support systems can help reduce personal distress – by decreasing activation in the dorsal anterior cingulate cortex region of the brain.[20]

PROCESSING LANGUAGE AND COMMUNICATION

In terms of social cognition, language is a significant improvement over hand gestures, facial expressions, and crude vocalizations. According to Dr. Lev Vygotsky, founder of cultural-historical psychology, culture provides us with a particular language and accompanying set of symbols. Vygotsky theorized that a child initially uses language to talk with others and also to him or herself, often aloud.[21] This "self-language" is gradually internalized, and along with other culturally-derived symbols, becomes a *psychological tool* that guides the child's thought processes, self-regulation, and behavior; a directing influence in the child's mind.[22]

While a baby's brain is able to respond to all human languages, languages are built on particular **phonemes** or units of sound. The phonemes that are repeated most become part of the brain's circuitry. By 10 to 12 months of age, a toddler's brain has started to distinguish among phonemes, remembering those of his/her native language and ignoring others.[23] The brain then learns to make words out of the sounds and attach meaning to the words. A full grasp of the language requires knowledge of sentence structure, tense use, and grammatical rules, and the ability to read. Children who are exposed to spoken language in their early years, who are read to and hear more, and more difficult words, find it easier to learn to read and comprehend what they read.[24]

Broca's area is the dominant region for speech, while **Wernicki's area** is the dominant region for language comprehension and meaning. Proper language function however, requires the integration of both hemispheres of the brain. While the left hemisphere houses the brain's language centers, processing the emotional content of language appears to occur in the right hemisphere. This means *communication* requires both hemispheres. Just as language is not the same as communication, speaking is not the same as writing. Speakers know who the receiver is and receive continuous feedback; they spend less time planning what they're going to say, and can lose sight of what's already been said; and speaking tends to be more spontaneous and less conscious than writing.[25]

When we engage in a conversation, we generally assume we share a certain amount of common knowledge and experience with the receiver[26]; an assumption that often leads to *mis*-communication. We tend to use the same words and expressions repeatedly (we "preformulate"), employ discourse markers to punctuate and guide the conversation,[27] and use intonation, cadence, and gestures – even when we're on the phone – to get our message across. We are also inclined to depend on **reflexive social language** to keep the conversation going and conflict free,[28] while the **internal dialogue** running inside our head is also largely reflexive. Finally, it takes a special kind of language, "the language of **self-reflection**," to stand back and think about what the conversation means and *choose* how to respond.[29]

PROCESSING...

- <u>Rules</u>. When "the environment suggests a disregard for the rules," the spillover effects can be widespread.[30] If we see other people breaking the rules, "we are more likely to give in to *any* of our own

- impulses,"[31] even if our own weakness has nothing to do with what we witnessed.[32]
- <u>Deception</u>. Deception is thought to have survival value because it disguises our intentions and feelings. Even a four-year-old can be deceptive.[33] Research has found that our ability to detect deception is only slightly better than chance.[34] Moreover, as a largely unconscious process, when we consciously work to detect deception, we actually *decrease* our accuracy.[35] *Self*-deception may have also been naturally selected for because it helps us believe our own deception.[36]
- <u>Disgust</u>. Disgust is thought to be learned, based on the fact that it varies from culture to culture and babies have to be trained to find certain things, like feces and sweat, disgusting.[37] The disgust response increases activity in two regions of the brain, the **insula** and the amygdala.
- <u>Aggression</u>. When it comes to processing aggression, intention is crucial. Opinions vary as to whether aggression is inborn or learned; among primates, though, it is nearly universal. It's always easier to inflict pain on someone when we have dehumanized that person or group. And once we are aggressed against, our retaliation almost always exceeds the original offense.[38]
- <u>Revenge</u>. Revenge has been said to "sit comfortably and nobly with the select inventory of human instincts."[39] In his book, <u>Payback</u>, Thane Rosenbaum argues that the call for justice is always a call for revenge.[40] He suggests that revenge is part and parcel of our innate sense of fairness, and governed by the regions of the brain that process emotions and rewards.[41] He states: "The human brain reacts positively when it experiences the satisfaction of getting even...Revenge justly owed and justly taken... is one of the ways human beings demonstrate their commitment to moral order."[42]

A
Closing Thought

A Neuroscience Question: Is it neurons all the way...up?

We know there is no "little man" at the controls of the brain, and it is neurons "all the way down." As social neuroscience continues to shed more light on our social nature, one may wonder whether understanding neuroscience detracts or undermines our belief in God.

The literature seems to address this question in one of three ways. The first argues that what's been learned through neuroscience points to a system so perfect, how could one *not* believe in God? The second suggests that since science will never explain everything, there will always be room for God alongside science. The third says that the answer doesn't really matter; that we should focus on the things and people in our lives that are truly important.

Perhaps Dr. Robert Sapolsky in his Great Course on Biology and Human Behavior (2005) handled this question best when he said, "To explain something is not to destroy the capacity to be moved by it."

25

Theories about the Mind

So how *do* we make sense of another human being? The answer is **theory of mind**: we process their faces, words, body language, and then we process the other person's mind. Theory of mind (ToM) is the ability to reflect on one's self and understand the **mental states** of others – their beliefs, desires, and intentions. It assumes people act on knowable intentions and those intentions can be read by others with a certain degree of accuracy. We use our ToM to decide how other people feel and what they're like; what they're doing now and might do next; whether they like us and what they want; what they're thinking, needing, meaning, etc. We even use ToM to explain *why* they are as they are. We draw on everything we know about how the world works and come to a conclusion about another person's inner, unobservable, totally private, and directly unknowable mental state. And we do it intuitively, in a matter of seconds.

Theory of mind (ToM) can be thought of as a "web of assumptions/presumptions that enables large societies to function effectively."[1] It is a natural by-product of a brain that's becoming increasingly social, craves certainty, and needs to know. ToM:

- Is a type of cognition, influenced by information "inferred from the person, the environment, or a third source such as prior experience"[2];
- Serves as a useful survival skill: it pays to know what another person needs or expects of you as well as what he's up to;

- Reflects the recognition that to understand another person, we often need to go beyond words and into minds;
- Can be automatic or controlled, and yet the assessments we make of others and our subsequent responses are often unconscious and habitual;
- Is informed by culture, something we all think we're good at, and readily subject to error; and,
- Illustrates the extent to which our experience of another person is, at best, *indirect* -- an interaction of mental states and our interpretation of those states.

THEORY OF MIND AND THE BRAIN

The term "theory of mind," also known as **mentalizing**, was first coined in the mid-1970's by two researchers, David Premack and Guy Woodruff, who were studying whether chimpanzees could understand the mental states of other chimpanzees. In humans, theory of mind is thought to be supported by a "distributed network of brain regions" in which "different regions [likely] play different roles."[3] The most frequently cited regions include the **medial prefrontal cortex** (MPFC), **the temporoparietal junction** (TPJ), the **precuneus/posterior cingulate** (P/PC), and the **temporal poles** (TP). Working together, activity in the medial prefrontal cortex forms impressions about the internal states of others[4]; the temporoparietal junction is believed to play a "more general role" in reasoning about people's intentions (especially in the right temporoparietal junction)[5] and representing those intentions[6]; and the posterior and superior areas of the temporoparietal junction are thought to be associated with personal knowledge. The **posterior superior temporal sulcus** (pSTS), which is also often mentioned in theory of mind literature, focuses on discerning head position and other "biological motion"[7] related to a person's body language, and the temporal poles and precuneus are particularly engaged when reading stories which involve "mental state attributions…" and other similar activities.[8]

Theory of mind (ToM) is part of the brain's social survival tool kit, which also includes **self-awareness** to monitor our behavior and **self-regulation** to change our behavior in accordance with group norms and experience. Interestingly, ToM of mind may be an **exaptation**: it may have evolved from several "pre-existing neural networks" that helped us distinguish between self and other, etc., that were later "co-opted" for a different purpose such as "imitation, anticipation, and empathy."[9] We sometimes use our ToM to interpret or explain our *own mind*: self-knowledge can be thought of as "turning our mind-reading abilities on ourselves."[10] Doing so may rely on the same "sensory channels" and many of the same "sensory cues" we use to read others, making it "just as interpretive…as knowledge of the thoughts of others."[11]

One key to successful mentalizing is inhibiting our own perspective. We also need a brain able to keep track of relevant environmental cues, which has sufficient short-term memory to maintain content information, and has a long-term memory system capable of storing relevant ToM knowledge from past experience.[12] While **social thinking** and **non-social thinking** may *feel* like the same kind of reasoning process, they rely on distinct neural systems,[13] such that it's hard to do both simultaneously. ToM is not the same as **perspective-taking**: ToM can be either spontaneous or controlled; perspective-taking is a "more conscious form of mentalizing"[14] where we deliberately quiet our own perspective and consciously construct the perspective of another. Finally, for the development of ToM to occur on a "normal" timetable, "language acquisition and comprehensible communication from infancy are essential."[15] Studies have shown that "native-signing deaf children and typically-hearing control children performed at an equivalent level to each other" on both standard verbal ToM tests and the "low-verbal analogs" of ToM tasks.[16]

DEVELOPING A THEORY OF MIND

Theory of Mind is predicated on the belief that we each have a mind, and therefore there must be a reason behind our actions. Children have to *develop* that belief and the thinking processes and reasoning skills that go along with it. They have to *build* a theory of mind, step-by-step, beginning with the conceptual doors discussed in Chapter 12: among them, learning the difference between self and other; understanding there is a mental and a physical world; and recognizing that people have separate minds and hence, different thoughts. Developing a theory of mind (ToM) is a gradual process that begins at birth and lasts a lifetime. At the most basic level, a child must learn what a mind is and what it does.[17] Many of the building blocks of a ToM we share in common with our "ape relatives."[18] Valerie E. Stone describes three such building blocks:

1. *Learning the relationship between actions and goals.* Between 5 and 9 months of age, infants can distinguish between intentional and accidental actions, and by 15 months, they can "classify actions according to goals."[19]
2. *Joint Attention.* By 6 months of age, infants can interact with another person or an object. Between 9 to 12 months, they can follow a person's gaze, the direction of a finger pointing, and imitate certain acts.[20] Full joint attention appears around 18 to 24 months when the child starts to see others as "intentional agents like the self."[21]
3. *Pretend Play.* Around 18 to 24 months, children start to pretend play and learn to "decouple pretend reality" from "perceptual reality."[22] While they are still unable to understand their playmate's mental states, children do perceive play as something special,[23] and perhaps different.

Between two and three years of age, children begin to grasp that mental and physical things have different "properties" and "that mental states such as desire and knowledge are private, internal, and can change independent of reality."[24] This is the beginning of mentalizing. *Just as in adults*, the child is "less aware of the mind than of the self" and is thinking in terms of "me" or "him," not "my mind" or "his mind."[25] Around 2 years of age, children begin to show a "genuine reference to the subjective mental state of desire."[26] They learn that if someone feels a given way there must be a reason, and that one can tell how someone feels from his/her appearance or behavior.[27] In other words, they start to *infer* from the behavior of others. By about 4 years of age, children have developed what researchers call a **representational theory of mind**: they understand that: (i) "the mind is active, that it construes and interprets situations"; (ii) "mental entities" are "representations produced by the mind"; and, (iii) people "take those representations to be truly the way the world is."[28] Thus by age 4, at a very preliminary level, children understand that we create a reality, and live in and through it.

Theory of mind research uses a variety of methods to test for the presence of theory of mind in children, the most famous being the **Sally/Anne False Belief task**. Children are shown pictures of two children in a room, Sally and Anne. Sally has a marble and puts the marble in a basket. Sally then leaves the room. While she's gone, Anne takes the marble out of the basket and puts in into a box. When Sally comes back into the room, the question for the child is this: "Where will Sally look for the marble?" The child must decide whether Sally will look for the marble in the basket (where she left it), or in the box (where it is). A false belief is when "the contents of the world" are seen to "contradict the contents of thought."[29] It's not until 4 years of age or older that a child can suppress what he/she knows (that the marble is in the box) and say that Sally will look for the marble in the basket where she left it. A child is said to have a theory of mind when he/she can recognize the consequences of having a false belief.[30]

Theory of mind development doesn't end at 6 or 7 years of age. It grows in complexity, becoming "progressively integrated with language… and knowledge…Some of these abilities are then automatized into efficient, but inflexible routines…[Many] remain largely intact into adulthood."[31] By preadolescence, a test for theory of mind examines four aspects of reasoning: the ability to understand multiple perspectives, recognize and understand emotional states, understand the other person as a psychological being with stable personality characteristics, and imagine multiple perspectives and alternatives.[32]

HOW WE READ A PERSON'S MIND

A debate exists in social neuroscience over *how* we read another's mind. One group maintains that we generate a representation of the other person's mental state in our brain and infer from it what they're thinking, feeling, or desiring. Another argues that we simulate another's experience in our own body and mind, and proceed based on what we're experiencing.

"Theory of mind-proper" is based on **representations** and derived through **inference**; as we said, the child develops a *representational* theory of mind. It is theorized that as we grow, we accumulate a mental library of representations of mental states from past experience. When we perceive another person, we call up the representation that best matches what we're seeing and infer from it the beliefs, intentions, thoughts, and desires of that person. In contrast, the simulation model, referred to as **Simulation Theory**, suggests that we "create within ourselves a simulation of the others' motor acts" and from these, "estimate and predict the mental processes of others."[33] It theorizes that we *feel* what we believe to be their thoughts, desires, and intentions. **Mirror neurons** lie at the heart of the Simulation Theory.

A. The Mirror System

Prior to the mid-1990's, neuroscientists believed the brain had separate regions for **perception** and **action**.[34] While studying macaque monkeys, Giacomo Rizzolatti and his colleagues discovered neurons that "were activated when the subject performs a given motor action, as well as when it observes another subject perform the same action."[35] In 1999, researcher Marco Iacoboni found regions in the human brain had similar "mirror properties."[36] Mirror neurons are **motor neurons**. They are found in humans in "a distributed network of neural regions" including the **premotor cortex, parietal lobes**, and **temporal lobes**.[37] Because fMRI (functional magnetic resonance imaging) does not examine individual neurons, in humans these regions are said to comprise **the mirror system**, as opposed to the mirror *neuron* system.[38]

Mirror neurons and the mirror system perform both motor *and* perceptual functions. The fact that perception and action can occur in the same neuron means we can imitate the actions of others. Moreover, we can do so *towards a specific end or purpose*: If I see you reach for an apple and bite into it, I can do the same and expect to have the same satisfied feeling you're showing on your face. The mirror system "makes it possible for us to understand the actions of others as if we were engaging in the action ourselves."[39] It also allows us to predict those actions: since actions tend to happen in sequences, mirror systems "seem to trigger" the representation of actions that will come "in the next hundreds of milliseconds."[40] In addition, many of our *reflexive* **resonance behaviors** are thought to be based on mirror neurons.[41]

More than discerning or mimicking an action, however, many simulation researchers believe the mirror system is how we understand people's *inner* states. They argue that mirror neurons "are sensitive to the *abstract meaning* of other people's actions"[42] (emphasis added) – they enable us to appreciate "what and how others think, feel, and experience,"[43] "adopt another person's conceptual vantage point," and link distinct inputs in abstract ways.[44] Marco Iacoboni summed it up this way: mirror neurons "code intentions" and hence, "clearly support" the simulation model of how we read other people's minds.[45]

B. Theory of Mind or Simulation Theory, Mentalizing or Mirroring?

There are thus two theories for how we make sense of another person's mind. Simulation Theory is based on the mirror system: it "plays a key role in perceiving what others do and how they are doing it by triggering an internal simulation of perceived actions."[46] Theory of mind (ToM) is based on representations from which we draw inferences: we generate a representation in our own brain, one that arises from past experience, and use it to infer another person's mental state. ToM is sometimes called "theory-theory" because our representations are theories about other people's minds and we use them to infer from one situation to the next; also, ToM is sometimes said to include "abstract general rules or laws" about how the mind works.[47] The false belief test is a good illustration of representational ToM. Simulation Theory, on the other hand, is seen as a more **embodied** form of social cognition: we use our mind as a proxy, putting ourselves in the position of another and imagining what it's like to "be them."[48]

To add to the confusion, multiple terms and definitions are bandied about in different texts. Theory of mind sometimes refers to everyday reasoning about the world, mind, and behavior; imputing invisible mental states or one's own internal states onto others; or *logically* inferring another's mental state. Simulation can mean understanding the goal of observed motor acts; making sense of another's behavior as if it was our own; or vicariously experiencing other's internal states. Mind-reading can mean the ability to stand in another person's shoes, and mentalizing can refer to *consciously* attributing the mental states of others.

C. Two Distinct Systems

Theory of mind and simulation are supported by two distinct *neural* systems. They have distinct functions as well. A variety of methods have been proposed for how to these two neural systems might work together: they may be employed concurrently; there may be high and low components of understanding another's mind; different situations may call for different social cognition tools, etc.[49] The biggest drawback of the mirror system is that it explains how we understand *actions*. What we know about actions may not apply to feelings and beliefs, especially ambiguous ones.[50] Matthew Lieberman, author of Social, suggests that the two systems operate as "opposing neural systems": when one system is activated, there is less activity in the other.[51] That means the more we think about people, the less we mirror them. The relationship between these two systems may therefore be complementary, with both being essential for reading others' minds.[52] According to Lieberman, Simulation Theory and the mirror system focus on the lower-level *motor* intentions of others, while theory of mind and the mentalizing system attend to higher-level intentions.[53]

AN ART NOT A SCIENCE

While the debate continues over precisely how we read another person's mind, we are nonetheless overconfident in our abilities.[54] Like any social cognition skill, however, reading others is prone to error and there are many reasons for this:

- We have a "pre-potent desire" to impose our mental states and knowledge onto the person we are perceiving,[55] and we make mistakes when we use our mind as a proxy for other minds[56];
- Language can confuse, influence, or sway our sense of another person. People can also deliberately try to deceive us, or simply say one thing and mean another;
- When we read another person's mind, it requires that we know ourselves which, as we've seen, we're not especially good at; and,
- While we may be able to quiet our viewpoint to some extent, we readily apply stereotypes, use heuristics, and take other shortcuts to avoid having to do the hard work of understanding another person.

Reading others' minds is best thought of as a baseline competency, requiring practice and effort to improve it. Nicholas Epley, author of Mindwise, points out that the longer or more intimately we know someone, the more inclined we are to assume we can read them accurately; what he calls the "illusion of insight."[57] In one study, the difference between "how much partners *actually* knew about each other and how much they *believed* they knew" grew *larger* the longer the couples had been together.[58] In sum, whether we're inferring or simulating, we're still conjecturing about how someone else feels, about their directly unknowable thoughts, intentions, needs, or desires. It's an *interpretation*. It's an estimate…a guess. This is one of the reasons why the **wisdom** of the brain points to the **importance of listening to other people's stories – of being with someone without**

needing to fix anything, so we can make the best use of the capacities we have.

Recommended Reading
Social, by Matthew D. Lieberman

26

Theory of Mind in Action: Empathy

Relationships provide us with our greatest opportunity to come to know ourselves. In our relationships, empathy is an opportunity to feel, to "participate in another's feelings or ideas." As a subset of theory of mind (ToM), empathy is ToM in action because in order to participate, we must first understand what a person is feeling. Empathy involves *affective* understanding. It is an essential component of helping and perhaps, one of the reasons we became helping professionals in the first place.

A PROBLEM OF DEFINITIONS

To study empathy, it might be useful to first define it. Empathy can be:

- "A complex form of psychological inference that combines observation, memory, knowledge, and reasoning"[1];
- "An affective response" that comes from understanding "another's emotional state or condition, and which is similar to what the other person is feeling or would be expected to feel"[2];
- "To imagine oneself in the other's situation and to experience, to some degree, the emotions that the other is experiencing"[3]; and/or,
- "A sense of knowing the personal experience of another (Goubert et al., 2005), a cognitive appreciation that is accompanied by both affective and behavioral responses."[4]

The word empathy has its origins in Greek, meaning "to suffer with."[5] According to the literature however, there are at least eight "related but distinct" uses of the term[6]:

1. "Knowing another person's internal state," especially his/her thoughts, feelings, etc.;
2. "Adopting the posture or matching the neural responses of an observed other," as in motor mimicry or matching another's neural representations;
3. "Coming to feel as another person feels," meaning to have a shared sense of feelings or physiology;
4. "Intuiting or projecting oneself into another's situation";
5. "Imagining how another person is thinking and feeling," such as taking another person's perspective;
6. "Imagining how one would think and feel in the other's place";
7. "Feeling distress at witnessing another person's suffering"; and finally,
8. "Feeling for another person who is suffering" which is also called pity, compassion, or sympathy.

Surprisingly, only one of the eight definitions involves *taking action* on the feeling of empathy. One author states, "most clinical and counseling psychologists agree" that *true* empathy requires three separate skills: being cognitively able to "intuit" another person's feelings, being able to emotionally share in those feelings, and having a "socially beneficial" *intention* to respond in a compassionate way.[7] Another author suggests that *full* empathetic understanding requires not only matching another's affect, but also producing the *urge* to help.[8] Neuroscientists are said to define empathy as a "complex form of psychological inference that enables us to understand the personal experiences of another person through cognitive, evaluative and affective processes."[9]

Generally speaking, most definitions of empathy "are more about understanding and perspective than about being driven to help another reduce suffering."[10] In fact, much of the research on the neuroscience of empathy has focused on **affect-matching**, and in particular, on the "neural basis of affect-matching" in the area of pain and pain distress.[11] Social neuroscience is only now starting to examine "how empathy might lead... to helping."[12]

THE NEURAL BASES OF EMPATHY

Empathy involves mind-reading and mentalizing; shared feelings and modeling; feeling another's pain or distress; memories, general knowledge, reasoning skills, and inference; imagination, perspective-taking, role taking, and mimicry; and potentially, motivation -- the intention to act and/or take action. It can be thought of as a *hypothesis* we form about another person based on visceral, cognitive, and emotional forms of information.[13] Not surprisingly, it relies on multiple neural networks including, but not limited to:

- Those that support theory of mind, including the **medial prefrontal cortex** (MPFC), the **temporoparietal junction** (TPJ), the precuneus/**posterior cingulate** (P/PC), and the **temporal poles** (TP);
- The **mirror system**;
- The **pain distress system** including the **dorsal anterior cingulate cortex** (dACC) and the anterior **insula**;
- Those that support emotion including the **amygdala** and the myriad projections to and from it;
- The **sympathetic and parasympathetic nervous systems, the smart vagus, and the somatosensory cortex;** and,
- The medial prefrontal cortex as well as the **orbitofrontal** (OFC) and **orbitomedial prefrontal cortex** (OMPFC), both of which are involved in emotions and emotional regulation.

In terms of how these and related networks work together to support empathy, two areas of the temporal cortex -- **the fusiform face area** and the **superior temporal sulcus** (STS) -- help decipher "biological movement and intention" and in so doing, set the stage for us to feel safe enough to be with another person and not react defensively.[14] While the medial prefrontal cortex is activated when we are thinking about both self and other, the **ventral medial prefrontal cortex** (vMPFC) is especially activated when thinking about a person similar to us.[15] This is important because our capacity to empathize often depends on who the other person is and how we perceive them.

fMRI research suggests that different "nodes" of the theory of mind network are engaged when we cooperate versus compete with others: the posterior cingulate cortex is more active when we cooperate, the medial prefrontal cortex during competition.[16] Research has also found that the core network involved in empathy for pain is present when we feel disgust, embarrassment, or react to social exclusion.[17] Another important region is the **septal area**. With respect to empathy, it is thought that information "converges" in the septal area which "converts" that information into the "urge" to act.[18] According to animal research, the septal area is involved in processing rewards, down-regulating fear, and maternal caregiving.[19] As we've said, we are driven by both social and self-interest – in this case, to feel safe *and* rewarded.

EMOTIONAL AND COGNITIVE EMPATHY

Empathy has many benefits. It promotes social understanding, pro-social behaviors, compassion and caring, and creates a feeling of mutual sensitivity, perhaps even vulnerability. It's still not clear whether when we help others we do so because "of a selfless empathic concern" for them, or because we feel similar and somehow connected to them, and hence feel their suffering.[20]

Recent evidence supports "a model of two separate brain systems for empathy: an emotional system and a cognitive system."[21] **Emotional empathy** involves using our own internal mechanisms to estimate and predict how another person feels. It is closely tied to **Simulation Theory**, and **the mirror system** which has been called the "root of empathy."[22] In Simulation Theory, we "automatically create within ourselves a simulation" of the other person's "motor acts," and consequently "feel the desires, preferences, and beliefs of the sort we assume the other to have."[23] Emotional empathy involves what's called affect- or **state-matching**: it's thought that when we perceive or simulate the emotions of another, the regions of our brain associated with those emotions are activated such that we "match" their feeling state, and feel as they are feeling. Emotional empathy is related to **imitation** or mimicry, the notion of emotional contagion, and the idea of "shared pain" in which we "feel the pain" of another on some level.

Cognitive empathy involves seeing the world from another person's perspective. It's thought to require "higher-order cognitive functions,"[24] and is closely linked to **Theory of mind** and **inference**. According to Theory of Mind, we create a picture of another person's mental state by selecting among various representations of mental states in our own mind, and "cognitively take the perspective" of the other person.[25] Cognitive empathy is related to **perspective-taking**, where we must consciously inhibit our first-person perspective and construct the views of another person; also, to abstract reasoning and cognitive flexibility.[26] It's likely that the *full* experience of empathy requires *both* the emotional and the cognitive empathy brain systems be activated.[27] It is still, however, a guess, a hypothesis about how another person feels: a simulation is never one-to-one and an inference is still a construction.

THINKING ABOUT EMPATHY

We know that empathy is a skill that must be worked on. Even therapists have to practice "de-centering from their own experience," tuning into their bodies, and "listening carefully to the details" of their clients' experiences.[28] In addition:

- Our willingness and ability to empathize with others is strongly influenced by who they are and how they behave.[29] It's easier to empathize with a loved one than a stranger, with in-group than with out-group members, and for a victim who is close to us or someone we care about[30];

- Some studies report that females are more empathic than males.[31] Others suggest that men and women have about the same empathic ability. When faced with "situational cues" that remind women they are *supposed* to be more empathic than men, women can "outperform men on empathic accuracy tasks"[32];
- Projecting our knowledge and feelings onto others is a common problem in empathy. Studies have shown that people are "likely to project not only what they know, but also what they erroneously believe they know." The level of confidence one has in his/her own knowledge "was consistently a strong predictor of the probability of projecting"[33]; and,
- In terms of empathizing with another's pain, research has found that our response can be either self- or other-oriented. An "other-oriented" response can lead to a "full understanding," whereas a self-oriented response (called **personal distress**) can cause one to focus on relieving one's *own* distress.[34] Mitigating factors include the intensity of the pain, the "affective link" between the observer and the target, and whether the pain was perceived as a "justifiable cure or not."[35]

Finally, contextual information and individual differences affect how accurately we empathize with others. When people received immediate feedback about the other person's *actual* thoughts or feelings, they "subsequently displayed better empathic accuracy."[36] When *verbal* information was excluded, empathic accuracy was "dramatically impaired," while the loss of visual information had "a surprisingly negligible effect."[37] Among married couples, studies have shown that beginning after the first or second year of marriage, empathic accuracy "significantly declines."[38] Overall -- comparing the degree to which the "thought-feeling events" reported by the target match those inferred by the observer -- "the data from many studies show that the typical range of mean empathic accuracy scores is from 15% to 35%."[39]

IN SUM

Empathy takes time. It points to the final piece of **wisdom** that the brain has to offer in this book. Empathy invites us **to experience what lies beneath our busyness** -- when we're not absorbed, or rushed, or deep in thought. What lies beneath our busy mind? It's possible that there rests our capacity for empathy…to suffer with and act on it.

Figure I
**Summary of the
Key Principles of Brain Operation**

Key Principles	Characteristic Human Traits
<u>Neurons connect with other neurons to form pathways of communication</u>	• Our constant state of vigilance and the primacy of self-preservation • Human behavior as being need-driven • The critical link between emotion and the viscera, and healing • The highly generalizable nature of fear • The need to organize life around a self • A default system that focuses on self/others • How a change in physiology alters behavior • The emphasis on repetition, practice, and rehearsal • Our basic social nature
<u>The brain is integrative; it works together as a whole</u>	• The way networks form around a task, dissolve and reconfigure as situations change • The dependency of complex skills on many distinct interconnected cortical and subcortical regions • How the brain constructs the perception of reality; of color, sound, smells, etc. • The simultaneous processing of sensory and motor information, and that cognition and emotion are inextricable • How the brain-body-environment interaction creates a single, unified experience

The brain naturally looks for patterns	• How patterns of neural connections form recurring thoughts, emotions, and behaviors • The way information becomes representations, and is linked together to form ideas and action plans • How the brain functions as an "association maker" and a "rule maker" • The brain's preference for certainty and search for predictability • The way "knowing" creates a self-reinforcing feeling of rightness and conviction • How *every experience* is a projection of "me"
The brain learns from experience	• The key role adaptability plays in how the brain seeks to manage the environment • That we fight when angry, flee when scared, and freeze when the situation is lethal • The brain's capacity for plasticity…within reason • Our ability to feel safe, okay, and capable • How self-regulation is experience-dependent • Our capacity to learn throughout life so long as there is something to learn • How new learning attaches to our established base of knowledge

The brain learns and remembers what has meaning	• The integral nature of meaning to our personal narrative, the story of "me" • How the past acts as "synaptic shadows" for future reference • That the content of memory is unique to each individual and a source of individuality • How the past becomes the filter through which all information is processed • How beliefs and belief systems are learned • That memory can and often does fail • How a threat to our meaning systems can feel like a threat to our very sense of self • That the left hemisphere interpreter needs to know and explain why
The brain makes as much as possible automatic and unconscious	• How reflexes, procedural memory, our sense of self, and our own reactivity happen before we are consciously aware • How implicit memory creates mind patterns that frame our entire lives • The natural tendency to revert to the path of least effort • The use of stereotypes, habits, and other short-cuts to organize and simplify a complex social world • The fact that more and increasingly social faculties are capable of being produced unconsciously • How we automatically adapt to a particular culture and abide by its norms • The fact that consciousness has to be engaged

We are, and hence our brain is, increasingly social	- That the brain processes social pain in much the same way as physical pain
- The fact that we imitate, simulate, and learn from others
- That culture and relationships play a large role in defining our sense of self
- How ToM serves as a survival tool, and develops over a lifetime
- The controlling nature of culture and groups, and prevalence of group mind
- That we use our mind as a proxy to interpret others' emotions and behavior
- How we find it difficult to think socially and non-socially at the same time
- That relationships are key to understanding ourselves

27

Neuroscience and the Future of the Helping Professions

Neuroscience offers us a deep look into human nature. It provides a clearer picture of human behavior. It answers *why*: why we defend ourselves, why we persevere, and why change is so difficult; why we sometimes resonate, others times erupt, and on occasion, calmly resolve a situation. What we do in the helping professions with what we learn from neuroscience depends on us. The wisdom of the brain could be pointing to new directions in what it means to help -- it's up to us to figure these out. There are five summary features of the brain that every person brings to every situation. Each inspires slow and steady growth in the ever-changing nature of the helping professions.

THE FIVE SUMMARY FEATURES AND WHERE THEY POINT

1. Perception is Constructed – In the Future, Help may Begin with the Body.
Perception is the product of the senses; it is constructed. There is no such thing as color or tone, smells or tastes outside the brain: the brain takes in electromagnetic waves, pressure waves, and chemical compounds and creates our perceptual experience. The brain can only interpret what it takes in within the confines of past experience. It must first deconstruct

the experience before putting it back together again and when it does, it adds in emotions, drives, and affective values; wants, desires, and dreams; and the rules, biases, and expectations we've collected over a lifetime. In this way, the brain constructs *both* our perception and our experience of reality: what we see, hear, feel, taste, and smell is filtered through the hidden layers of our unconscious; it's a projection of who and how we are.

The all-absorbing nature of our constructed perception speaks to the need to slow down and take a step back to observe what's happening. To support this, help in the future will likely **begin with the body** – in the future, the helping professions may increasingly attend to the embodied nature of the brain. Because our perception is constructed, it can be re-constructed from a different mind/body-set. The body is the easiest portal because it is the first place we experience new beginnings and healing; it's where we feel even the slightest change and see results the quickest. As a first step, it may be as simple as encouraging an angry person to try to stop eating spicy food for a month; an over-thinking person to consider taking an evening walk, alone and outside; or an opinionated person to sit and listen, without speaking. In each case, we observe what happens in the body.

Bottom-up measures like these help us attend to what we're experiencing *inside*, engaging the medial prefrontal cortex (MPFC) and its abundant subcortical connections.[1] Other similar measures might include undertaking a physical activity, getting some fresh air, using slow neutral music or soothing scents as a backdrop, or watching how emotions and states of arousal pass -- just like thoughts, they come and go. As professionals, we can also use our breath as a tool, taking a long, deep breath in and releasing it slowly. Nine times out of ten, others will follow: shoulders drop, jaws release, and tension decreases. Bottom-up measures are designed to get us out of our head and draw us back into the body. They help refocus thoughts and needs on the basics of living and being alive.

Beginning with the body feeds the *neuroception of safety*, the feeling of safety at the neural level. Over time, it may become commonplace for helping professionals to anticipate that the brain's first re-action will be one of survival; to recognize that emotions need expression because they are a form of communication; and to appreciate that our clients' subjective inner world is his or hers alone. We will come to look for all the ways people seek certainty -- how we strive to put a rug back under ourselves even after the last one failed; how we drift back to the path of least effort. Most importantly, we will be guided by the precept that safety is personal, because safety is how we discover our own *personal* sense of freedom.

2. Prefrontal Capabilities are Intentional – The Future Day-to-Day Work of the Helping Professions may emphasize Becoming Aware of Mind Patterns.

The prefrontal cortex (PFC) is abundantly connected, with projections to and from a vast array of other cortical and subcortical regions. It's the connectivity of the prefrontal cortex which makes its vast functionality possible. As a general rule, the dorsolateral prefrontal cortex focuses more on goals, planning, consciousness, and attention, while the orbitofrontal region is especially involved in emotional cognition and bodily awareness. The prefrontal cortex manages attention, supports working memory, oversees emotions, and integrates thoughts with emotions. It plays a key role in social interactions, imagining the future, comparing multiple contingencies over time, language, and dealing with novel or uncertain situations.[2]

The prefrontal cortex is more intentional than it is rational. Its capacity depends on the quality of early role models, the success of critical learning periods, the breadth and diversity of opportunities, genetics, temperament, etc. More than expecting it to assert control, the strength of the prefrontal cortex lies in its *generative* powers. In the near future, the day-to-day work of helping professionals may emphasize **bringing individual and collective mind patterns into conscious awareness** – to identify the associations, inferences, and causal relationships that frame our lives and discover how every experience is a projection of "me."

Because we are always involved with others, relationships provide us with all the material we need to examine our mind patterns. While the prefrontal cortex will generate many explanations to try to explain those patterns, some will resonate in the body/mind. If we're angry with someone, what is it that we are struggling with? Are we angry with aging, with the loss of time, with the if/then rules that didn't deliver, with God? If we can't tolerate someone's greediness, their over-eating, arrogance, or unwillingness to change, etc., where *in us* does that projection come from and what do we want to do with it? Our relationships serve as a mirror to reflect back to us the mind patterns that dominate our lives.

The helping professions have many top-down techniques for bringing mind patterns into the light; most complement those that work with the body. One of the best involves asking well-crafted, inquiring *but not probing* questions.[3] A well-crafted question is not purpose-driven but asked because we do not have the answer; it's elicitive – it *invites* the other person to pursue an answer. Elicitive questions might include:

- How does this have meaning for you?
- When your shoulders droop like that, what's going on for you?
- Is this familiar to you?
- What's happening now for you?
- What's at the heart of this for you?
- Is there something beneath that or around it that's important to you?
- Have you had to resolve something like this in the past?
- What might change for you if this was different?
- Is there a question you'd like me to ask you?

Well-crafted question encourage us to go inward which can support the orbitofrontal cortex in doing its job and sometimes, calm the amygdala. Once asked, such questions open space around and within both the asker and the asked. To accompany such questions, helping professionals may focus on *noticing* as a means to look past the single, unified whole of experience. *Noticing* "immediately helps…shift our perspective and opens up new options other than our automatic, habitual reactions."[4] We notice gently and without intent, providing space for whatever happens next. To bring mind patterns to light, we must accept that much of what we learn is not in our control and yet, we are accountable. We must also anticipate that change can happen...within reason; that some things are harder to change, and like us, people learn what *they* need.

3. Individual Differences are Significant – The Goal of the Helping Professions may be to Increase Self/Other Awareness.

No two brains are alike: cell structure, regions, connections, and the strength of those connections differ. Individuals have different genetic attributes, numbers of dendrites, amounts of and sensitivity to neurotransmitters, thresholds of neural excitability, patterns of connectivity, and uniquely different network topographies. Fetal experience, hormone levels, and pain receptors vary person-to-person; as do environmental sensitivity, the capacity for long-term potentiation (LTP), and brain plasticity. We each have a subjective inner milieu; our own internal mapping systems, emotional set-points, and tolerance for stress. We are individuals built through relationships, where we accumulate a unique past, dissimilar memories, and unconscious constructs; we live in a particular culture, internalize a certain level of control, and come equipped with a socialized sense of self. In sum, we have different brains *and* different minds. We will always make sense of the world, maintain a coherent sense of self, and sustain some semblance of control in a very personal way. We can therefore, at best, know the experience of another indirectly because our brain exerts such a powerful influence over how we construct reality.

To encourage the capacity to relate to everything, the main goal of future helping professions may be to **increase self/other awareness**. This will require we learn more about social cognition, our inherent social nature, the default network, theory of mind, sense of otherness, and capacity for empathy. As social neuroscience continues to gain ground, the diverse fields that comprise the helping professions may look for new ways to:

- Notice the voice inside our head -- the voice that wants to believe we are permanent, things are fixed, and others are scary. In time, we may even seek to silence that voice;
- Take note of the space we take up in a room – at a party, at the office, at home with the family. In effect, looking at how often and to what extent, we lead with our ego; and,
- Strive to ask, not tell.

One of the key skills will likely be *pause and reflect*. A pause slows things down: it's takes a temporary lull and turns it into a moment to go inward, to reflect internally. Pause and reflect gives the body a chance to breathe; for the parasympathetic nervous system to return the body to a state of calm and for the brain to ponder. In time, as we increase self/other awareness, we will grow more attuned to the effect we have on others; a skill set that has the greatest payoff for us as a collective of individuals.

4. We are Heavily Influenced by the Unconscious –The Process of the Helping Professions may center on that which is Emergent.

The brain seeks to make as much unconscious as possible. Doing so saves energy, generates efficiencies, and reserves conscious attention for when it is needed most. Some 95-98% of what goes on in the brain happens outside conscious awareness. Many behaviors begin as conscious before they are made automatic and become unconscious; many can occur both consciously and unconsciously. We also know the list of formerly conscious behaviors that can be executed unconsciously continues to grow, and that consciousness *itself* depends on unconscious processes.

The breadth of unconscious influences encourages the future of how we help to **focus on the emergent** – on the answers, issues, ideas, and thoughts that come up from the inside, in our clients and ourselves. We need to listen to the stories others tell us without needing to fix anything, and to continuously doubt our assumptions, all the while appreciating just how much living goes on inside our head. What is poised to emerge lies right beneath the surface; it's the source of potential and meaning. It resonates in the subjective inner milieu and already has the investment of the client. It may be sitting in working memory just outside conscious awareness but it is the next step, the next transition, the next realization or opportunity from the vantage point of that person. It emanates from the brain-body-environment interaction and involves choices; some good, others not so good. By focusing on what is emergent, we resist the urge to need to figure things out, to find and/or offer solutions. The emergent orients the helping professions towards what is possible and fruitful.

Working with the emergent decreases our reliance on the familiar and cultivates a relationship with the unknown. It's difficult and sometimes scary work. It challenges the brain's conservative nature and counters our natural rigidity. It invites us to let go of *control*: to ask what it is we're holding onto and see what happens if we release it. It might involve accepting something about ourselves, giving away something we're attached to, or being vulnerable with someone who threatens us. It can also mean *doing nothing* – to wait, not having an answer or a comment, and allowing what is to emerge.

5. We Lead with the Self – The Future of the Helping Professions may rest with Moving Beyond the Mind.

The self is our organizing principle; it's the center of our experience. It comes with a feeling of agency -- a belief in and the experience of being able to act -- and a personal history, the story of "me." Episodic memory situates our sense of self in a given place and time; it is a constructive process, not a reproductive one.[5] Our sense of self is vulnerable, highly attuned to the opinions of others, and defined by where we fit, by where we belong. The self lives in the mind, but is not real, nor is it conscious; it is an object of consciousness.[6] The self is a series of thoughts linked together over time; a pattern of neural connections, thoughts, and ideas, and *not* the "direct perception of an agent."[7] Nevertheless, we believe we have far greater control than we actually have, and are driven by both social and self-interest. And everything is a projection of the self, of who and how we are.

The pervasiveness of the self suggests that a possible future for the helping professions rests with **moving beyond the mind**. Beyond the mind means beyond knowing and the need to know. It asks us to see through to those places where differences don't matter, and to continuously update and refine our viewpoint, making opinions an iterative process. It requires a full examination of what lies beneath our "busyness," and invites us to test the notion that we are not the mind because it can be observed. It is second order change. Such an endeavor has the greatest payoff for the species because it challenges evolution to invest in *self- reflection and introspection*; allowing us to look at our relationship with the environment, other people, and ourselves. Importantly, moving beyond the mind helps reveal our unconscious assumptions about living -- about what the world owes us, where we fit, how life should work out – enabling us to move beyond them as well.

IN CLOSING

We said in the beginning that neuroscience will change how we see ourselves. It will add depth to every interaction, and alter how we relate to and what we expect of others. In the end, we find that it takes the conservative, fearful, self-protective, re-active, and all-knowing aspects of human nature and entices the helping professions to consider beginning with the body; to examine our projected world with the goal of increasing silence and noticing the effect we have on others; to let go of control and look to what's emerging; as we learn to be in the body but beyond the self. Neuroscience seems to have delivered on its promise.

Appendix A

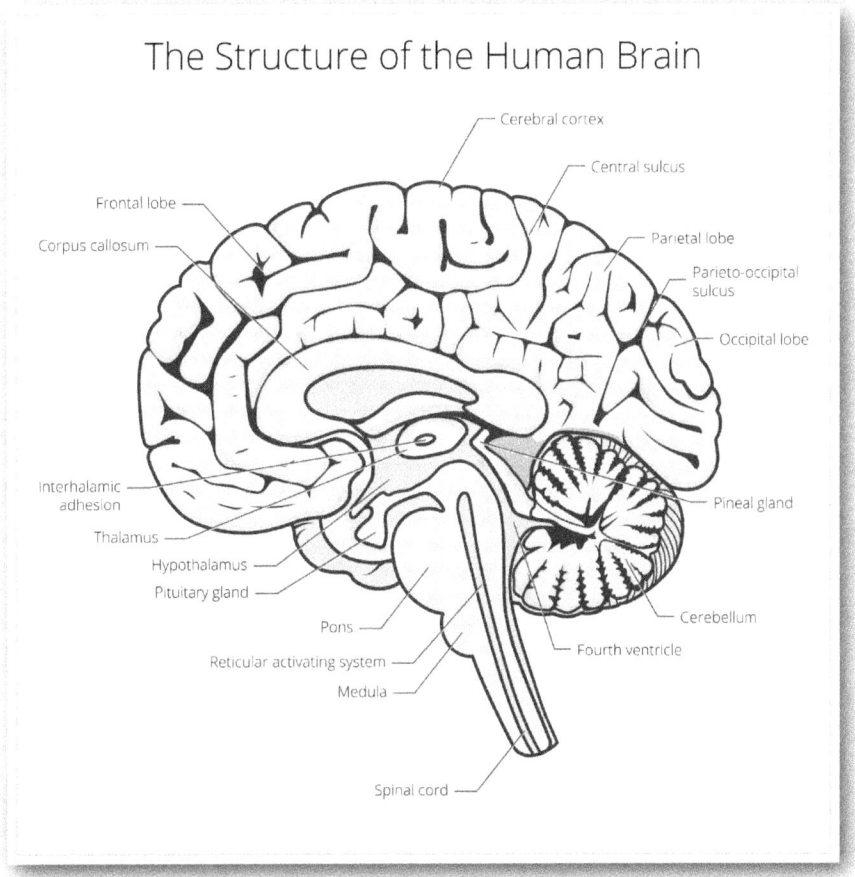

An Image of the Brain
Alexander Pokusay - Copyright 123RF.com

Author Biography

Camaron J. Thomas, PhD, received her doctorate in transpersonal psychology from Westbrook University and has spent a decade studying neuroscience.

Thomas is a professional mediator who specializes in family and community matters. She is trained in both transformative and interests-based mediation.

A certified integrated yoga teacher, Thomas has diplomat status in Ayurveda. She trained in mindfulness meditation under Dr. Jon Kabat-Zinn.

Thomas previously worked in public service, including as New York's first statewide director of technology. She has written three other books: Beyond Help, People Skills in Tough Times, and Managers, Part of the Problem?

Camaron lives with her husband and her dog, Charleston, outside Albany, New York.

References

PREFACE

1. Eric Kandel, et al. Principles of Neural Science, 5th edition, p. 1110. McGraw Medical, 2013.
2. NPR, Morning Edition, Today's Health. Interview with Wendy Wood, psychologist, University of Southern California. January, 2015.
3. Gollwitzer, Peter M. et al. *The Control of the Unwanted*, p. 500. The New Unconscious, Ibid.

CHAPTER 1

1. Gazzaniga, Michael S., Richard B. Ivry & George R. Mangun. Cognitive Neuroscience, 3rd edition, p. 87. W. W. Norton & Company, 2009.
2. Amthor, Frank. Neuroscience for Dummies, p. 231. Wiley & Sons, 2012.
3. Amthor, Frank. Neuroscience for Dummies, pp. 340-341. Wiley & Sons, 2012.
4. Amthor, Frank. Neuroscience for Dummies, Ibid., pp. 340-341.
5. Amthor, Frank. Neuroscience for Dummies, Ibid., p. 339.

CHAPTER 2

1. See: Gazzaniga, Michael S., Richard B. Ivry & George R. Mangun. Cognitive Neuroscience, 3rd edition, pp. 4-16. W. W. Norton & Company, 2009. See also: Kandel, Eric R. et al., Principles of Neural Science, 5th edition, pp. 5-18. McGraw Hill Medical, 2013.

2. The somatosensory cortex and its extensive connections, is involved in touch, pain, body temperature, and the position of limbs.
3. Gazzaniga Michael S., Richard B. Ivry & George R. Mangun. Cognitive Neuroscience, 3rd edition, p. 74. W. W. Norton & Company, 2009.
4. Gazzaniga Michael S., Richard B. Ivry & George R. Mangun. Cognitive Neuroscience, 3rd edition, Ibid., p. 4.
5. Amthor, Frank. Neuroscience for Dummies, p. 244. Wiley & Sons, 2012. See also: Gazzaniga, Michael S., Richard B. Ivry & George R. Mangun. Cognitive Neuroscience, 3rd edition, p. 648. W.W. Norton & Company, 2009.
6. Kandel, Eric R. In Search of Memory, p. 235 quoting Francois Jacob. W.W. Norton & Company, 2006.
7. LeDoux, Joseph. The Emotional Brain, p. 123 citing S.J. Gould (1977) and Pinker (1994). Simon & Schuster, 1996.
8. Allen, John. The Lives of the Brain, p. 27. The Belnap Press of Harvard University Press, 2009.
9. Hampton, Simon. Essential Evolutionary Psychology, pp. 42-55. Sage, 2010. See also: Kellogg, Ronald. The Making of the Mind, pp. 22-25. Prometheus Books, 2013.
10. Researchers caution us not to be too strict in our interpretation of "hunter-gatherer" or to read too much "sharing" into the word "co-operative" (See: Hampton, Simon. Essential Evolutionary Psychology, p. 54. Sage, 2010.)
11. Lawrence, Paul R. & Nitin Nohria. Driven, How Human Nature Shapes Our Choices, pp. 28-29, quoting Loewnstein. Jossey-Bass, 2002.
12. Renfrew, Colin. Prehistory, The Making of the Human Mind, pp. 67-69; 96. Modern Library Chronicles Book, 2007.
13. Renfrew, Colin. Prehistory, The Making of the Human Mind, Ibid, pp. 70-72, 115-129.
14. Gazzaniga, Michael S., Richard B. Ivry & George R. Mangun. Cognitive Neuroscience, 3rd edition, p. 635. W. W. Norton & Company, 2009.
15. Gazzaniga, Michael S., Richard B. Ivry & George R. Mangun. Cognitive Neuroscience, 3rd edition, Ibid., p. 649.

16. Gazzaniga, Michael S., Richard B. Ivry & George R. Mangun. Cognitive Neuroscience, 3rd edition, Ibid., p. 649.
17. Gazzaniga, Michael S., Richard B. Ivry & George R. Mangun. Cognitive Neuroscience, 3rd edition, Ibid., p. 662 citing Hutsler & Galuske (2003).
18. Kellogg, Ronald T. The Making of the Mind, Ibid., pp. 11-15.
19. Satel, Sally & Scott O. Lilienfel. Brainwashed: The Seductive Appeal of Mindless Neuroscience, pp. 1-21. Basic Books, 2013.
20. Functional specialization refers to whether a region specializes in one or more functions.
21. Taken from a lecture at the Neuroscience Bootcamp, University of Pennsylvania, 2013.
22. This is a composite from many conference presentations.
23. Hemispheric Specialization refers to the brain's two hemispheres and whether some functions are assigned more to one hemisphere than the other. Functional specialization refers to brain regions and whether a region specializes in one or more functions.
24. Gazzaniga Michael S., Richard B. Ivry & George R. Mangun. Cognitive Neuroscience, 3rd edition, Ibid., pp. 458-459.

CHAPTER 3

1. Amthor, Frank. Neuroscience for Dummies, pp. 23; 29. Wiley & Sons, 2012.
2. Simpkins, C. Alexander & Annellen M. Simpkins. Neuroscience for Clinicians, pp. 111-121. Springer, 2013.
3. See: Eagleman, David. Incognito, pp. 86-96. Pantheon Books. 2011.
4. Eagleman, David. Incognito, p. 87. Pantheon Books. 2011.
5. See: Winston, Robert. Human Instinct. Bantam Books, 2003.
6. Lawrence, Paul & Nitin Nohria. Driven, How Human Nature Shapes Our Choices, pp. 44-52. Jossey-Bass, 2002.
7. Lawrence, Paul & Nitin Nohria. Driven, How Human Nature Shapes Our Choices, Ibid., p. 131.
8. Lawrence, Paul & Nitin Nohria. Driven, How Human Nature Shapes Our Choices, Ibid., pp. 131-132.

[9] Leary, Mark. Understanding the Mystery of Human Behavior, Lecture 2. Great Courses, The Teaching Company, 2012.
[10] Shermer, Michael. The Believing Brain, p. 167. Times Books, 2011. See also: Pinker, Steven. How the Mind Works, p. 427. W.W. Norton & Company, 2009.
[11] See: Panksepp, Jaak. Affective Neuroscience. Oxford University Press, 2005.
[12] Panksepp, Jaak. Affective Neuroscience, pp. 188-189. Oxford University Press, 2005.
[13] Panksepp, Jaak. Affective Neuroscience, Ibid., pp. 212-213.
[14] Panksepp, Jaak. Affective Neuroscience, Ibid., pp. 249, 262-265.
[15] Panksepp, Jaak. Affective Neuroscience, Ibid., p. 191.
[16] See: Kahneman, Daniel. Thinking, Fast and Slow. Farrar, Straus and Giroux, 2011.
[17] Kahneman, Daniel. Thinking, Fast and Slow, p. 35. Farrar, Straus and Giroux, 2011.
[18] Amthor, Frank. Neuroscience for Dummies, Ibid., p. 336.
[19] Sapolsky, Robert. Biology and Human Behavior: The Neurological Origins of Individuality, p. 128 in Course Guidebook. Great Courses, The Teaching Company, 2005.
[20] Sapolsky, Robert. Biology and Human Behavior: The Neurological Origins of Individuality, Ibid., p. 81 in Course Guidebook.

CHAPTER 4

[1] Seung, Sebastian. Connectome, p. 60. Houghton Mifflin Harcourt, 2012.
[2] Ullian, Erik et al. *Role for Glia in Synaptogenesis*, Department of Neurobiology, Stanford University School of Medicine. Wiley Online Library, 2004.
[3] Gazzaniga, Michael S., Richard B. Ivry & George R. Mangun. Cognitive Neuroscience, 3rd edition, p. 20. W.W. Norton & Company, 2009.
[4] Amthor, Frank. Neuroscience for Dummies, p. 13. Wiley & Sons, 2012.

5. Somatosensory means related touch, pain, temperature, and limb positioning.
6. The environment inside the neuron is less positive (or more negative) than the environment outside the neuron. The gradient is between Na+ (outside) and K+ (inside).
7. Kandel, Eric R. et al. Principles of Neural Science, 5th edition, pp. 148-149. McGraw Hill Medical, 2013.
8. In a nutshell: at rest, there is a higher concentration of Na+ outside the neuron and a higher concentration of K+ inside the neuron. An action potential is initiated when ions move across the cell membrane, thus changing the "charge separation." In other words, as Na+ flows across the membrane, it depolarizes the membrane which triggers the action potential. The outflow of K+ then re-polarizes the membrane and re-establishes the initial charge separation. It is the rapid depolarization and subsequent repolarization of the membrane in a "localized area" that constitutes the action potential. See: Kandel, Eric R. et al. Principles of Neural Science, 5th edition, p. 170. McGraw Hill Medical, 2013.
9. Sporns, Olaf. Networks of the Brain, pp. 157-158. MIT Press, 2011.
10. Buckner, Randy L., Jessica R. Andrews-Hanna & Daniel L. Schacter. *The Brain's Default Network, Anatomy, Function and Relevance to Disease*, p. 1. Ann. N.Y. Academy of Sciences 1124:1-38, 2008.
11. Sporns, Olaf. Networks of the Brain, Ibid., p. 176.
12. Buckner, Randy L., Jessica R. Andrews-Hanna, & Daniel L. Schacter. *The Brain's Default Network, Anatomy, Function and Relevance to Disease*. Ibid., p. 30.
13. Kandel, Eric R. et al. Principles of Neural Science, 5th edition, p. 1079. McGraw Hill Medical, 2013.
14. Kandel, Eric R. et al. Principles of Neural Science, 5th edition, Ibid., p. 1079.
15. Eagleman, David. Incognito, p. 49. Pantheon Books, 2011.
16. Kandel, Eric R. et al. Principles of Neural Science, 5th edition, Ibid., p. 385.

CHAPTER 5

[1] What follows is taken from a collection of texts. See: Kandel, Eric R. et al. Principles of Neural Science, 5th edition, pp. 337-365. McGraw Hill Medical, 2013. See also: Gazzaniga, Michael S., Richard B. Ivry & George R. Mangun, Cognitive Neuroscience, 3rd edition, pp. 62-87. W.W. Norton & Company, 2009.

[2] Koziol, Leonard, F. & Deborah Ely Budding. Subcortical Structures and Cognition, p. 125 citing Schmahmann & Pandya (1977). Springer, 2010.

[3] Koziol, Leonard, F. & Deborah Ely Budding. Subcortical Structures and Cognition, Ibid., pp. 128 citing Houk & Mugnaini (2003).

[4] See: Koziol, Leonard, F. & Deborah Ely Budding. Subcortical Structures and Cognition, pp. 128-157. Springer, 2010. See also: Cozolino, Louis. The Neuroscience of Psychotherapy, 2nd edition, pp. 7-8. W. W. Norton & Company, 2010.

[5] Amthor, Frank. Neuroscience for Dummies, pp. 233-235. Wiley & Sons, 2012.

[6] The OFC and OMPFC are both heavily involved in emotional processing. In this text, the OFC will be used predominately; both are referenced when appropriate, and the OMPFC will be used specifically with respect to attachment and social appraisal (See Cozolino, Louis. The Neuroscience of Psychotherapy, 2nd edition, pp. 126; 254. W.W. Norton & Company, 2010).

[7] Siegel, Daniel J. M.D. Pocket Guide to Interpersonal Neurobiology, p.13-3. W.W. Norton & Company, 2012.

[8] Amthor, Frank. Neuroscience for Dummies, Ibid., p. 32.

[9] Cozolino, Louis. The Neuroscience of Psychotherapy, 2nd edition, p. 229 citing Rilling et al. (2002). W.W. Norton & Company, 2010.

[10] Cozolino, Louis. The Neuroscience of Psychotherapy, 2nd edition, Ibid., p. 230 citing Bechara & Naqvi (2004).

[11] Kandel, Eric R. et al. Principles of Neural Science, 5th edition, pp. 345-346. McGraw Hill Medical, 2013.

[12] Excerpted from Kandel, Eric R. et al. Principles of Neural Science, 5th edition, Ibid., p. 338.

13. Kandel, Eric R. et al. Principles of Neural Science, 5th edition, Ibid., p. 1513.
14. Amthor, Frank. Personal correspondence. June, 2016.

CHAPTER 6

1. V.S. Ramachandran. The Tell-Tale Brain, p. 221. W.W. Norton & Company, 2011.
2. Gazzangia, Michael S., Richard B. Ivry & George R. Mangun. Cognitive Neuroscience, 3rd edition, p. 644. W.W. Norton & Company, 2009.
3. Gazzangia, Michael S., Richard B. Ivry & George R. Mangun. Cognitive Neuroscience, 3rd edition, Ibid., p. 643.
4. Amthor, Frank. Neuroscience for Dummies, pp. 227-228. Wiley & Sons, 2012.
5. Gazzangia, Michael S., Richard B. Ivry & George R. Mangun. Cognitive Neuroscience, 3rd edition, Ibid., p. 99.
6. Sporns, Olaf. Networks of the Brain, pp. 66-67. MIT Press, 2011.
7. Sporns, Olaf. Networks of the Brain, Ibid., p. 67.
8. Sporns, Olaf. Networks of the Brain, Ibid., p. 68 citing Prinz et al. (2004) and Marder & Goaillard (2006).
9. Amthor, Frank. Neuroscience for Dummies, Ibid., p. 33.
10. Depending on the text, the Insula and the Cingulate Cortex may be treated as part of the limbic system or as quasi-cortical lobes. The known role of each is growing in complexity.
11. Fuster, Joaquin M. The Prefrontal Cortex, 4th edition, pp. 30-31. Elsevier, 2008.
12. Fuster, Joaquin M. The Prefrontal Cortex, 4th edition, Ibid., pp. 34-35.
13. Cozolino, Louis. The Neuroscience of Psychotherapy, 2nd edition, p. 115. W.W. Norton & Company, 2010.
14. Amthor, Frank. Neuroscience for Dummies, Ibid., p. 165.
15. Amthor, Frank. Neuroscience for Dummies, Ibid., p. 165.
16. Amthor, Frank. Neuroscience for Dummies, Ibid., p. 167.
17. Satel, Sally & Scott O. Lilienfeld. Brainwashed, p. xix citing Steven Rose (2006). Basic Books, 2013.

18. See: Koziol, Leonard & Deborah Ely Budding. Subcortical Structures and Cognition. Springer, 2010.
19. Cozolino, Louis. The Neuroscience of Human Relationships, p. 60. W. W. Norton & Company, 2006.

CHAPTER 7

1. Kandel, Eric R. et al. Principles of Neural Science, 5th edition, pp. 333-334. McGraw Hill Medical, 2013.
2. Sporns, Olaf. The Networks of the Brain, p. 181. MIT Press, 2011.
3. Kandel, Eric R. et al. Principles of Neural Science, 5th edition, Ibid., p. 396.
4. Pinker, Steven. How the Mind Works, p. 26. W.W. Norton & Company, 2008.
5. See: Kandel, Eric R. et al. Principles of Neural Science, 5th edition, pp. 357-364, 375-377, 392-396, 510-518. McGraw Hill Medical, 2013.
6. Sporns, Olaf. The Networks of the Brain, Ibid., p. 72.
7. Sporns, Olaf. The Networks of the Brain, Ibid., p. 206.
8. See: Sporns, Olaf. The Networks of the Brain. MIT Press, 2011.
9. The terms "recursive," "reentrant," and "recurrent" are often, but not always, used interchangeably.
10. Sporns, Olaf. The Networks of the Brain, Ibid., p. 260.
11. Sporns, Olaf. The Networks of the Brain, Ibid., p. 150 citing Vogels et al. (2005).
12. Sporns, Olaf. The Networks of the Brain, Ibid., p. 152.
13. Sporns, Olaf. The Networks of the Brain, Ibid., p. 206.
14. Sprons, Olaf. The Networks of the Brain, Ibid., p. 180.
15. Sporns, Olaf. The Networks of the Brain, Ibid., p. 324.
16. Sporns, Olaf. The Networks of the Brain, Ibid., p. 193.
17. Wegner, Daniel M. *Who is the Controller of Controlled Processes?* p. 30. The New Unconscious, edited by Ran R. Hassin et al. Oxford University Press, 2005.
18. Wegner, Daniel M. *Who is the Controller of Controlled Processes?* Ibid., p. 30.

CHAPTER 8

1. Pinker, Steven. How the Mind Works, p. 26. W.W. Norton & Company, 2009.
2. Amthor, Frank. Neuroscience for Dummies, p. 335. Wiley & Sons, 2012.
3. Amthor, Frank. Neuroscience for Dummies, Ibid., p. 336.
4. Bor, Daniel. The Ravenous Brain, p. 22. Basic Books, 2012.
5. Bostrum, Nick. Superintelligence: Paths, Dangers, Strategies, p. 46. Oxford University Press, 2014.
6. See: Modell, Arnold H. Imagination and the Meaningful Brain. The Bradford Book, MIT Press, 2006.
7. Dozier, Rush W. Why We Hate, p. 101. Contemporary Books, 2002.
8. Panksepp, Jaak. Affective Neuroscience, p.144. Oxford University Press, 1998.
9. Panksepp, Jaak. Affective Neuroscience, Ibid., p. 161.
10. Panksepp, Jaak. Affective Neuroscience, Ibid., p. 162.
11. Amthor, Frank. Neuroscience for Dummies, pp. 174-175. Wiley & Sons, 2012.
12. Stress Effects and Stress Management, Psychological Harassment Information: www.psychologicalharassment.com/stress_and_stress_management.htm.
13. Van der Kolk, Bessel A. M.D. How the Body Keeps the Score, p. 65. Viking, 2014.
14. Van der Kolk, Bessel A. M.D. How the Body Keeps the Score, Ibid., p. 206 citing LeDoux (2000).
15. Burton, Robert A. M.D. On Being Certain, p. xiii. St. Martin's Griffin, 2008.
16. Burton, Robert A. M.D. On Being Certain, Ibid., p. 3, xii.

CHAPTER 9

1. Kahneman, Daniel. Thinking, Fast and Slow, p. 51. Farrar, Straus and Giroux, 2011.
2. Gazzaniga, Michael S., Richard B. Irvy & George R. Mangun. Cognitive Neuroscience, 3rd edition, p. 465. W.W. Norton & Company, 3rd edition, 2009.

3. Shermer, Michael. The Believing Brain, p. 87. Times Books, 2011.
4. Shermer, Michael. The Believing Brain, Ibid., p. 168.
5. Shermer, Michael. The Believing Brain, Ibid., p. 165.
6. Gazzaniga, Michael S., Richard B. Irvy & George R. Mangun. Cognitive Neuroscience, 3rd edition, Ibid., p. 379.
7. BBC – World Update, September 12, 2014.
8. Shermer, Michael. The Believing Brain, Ibid., p. 62 citing Foster and Kokko (2009).
9. Pinker, Steven. How the Mind Works, p. 307. W.W. Norton & Company, 2009.
10. Lilienfeld, Scott O. et al. 50 Great Myths of Popular Psychology, p. 61 citing Kastenbaum (2004). Wiley-Blackwell, 2011.
11. Eysenck, Michael W. & Mark T. Keane. Cognitive Psychology, A Student's Handbook, 6th edition, p. 56. Psychology Press, 2010.
12. The four lobes of the cortex are: Occipital, Parietal, Temporal, and Frontal.
13. Kandel, Eric R. et al. Principles of Neural Science, 5th edition, p. 564. McGraw Hill Medical, 2013.
14. Kandel, Eric R. et al. Principles of Neural Science, 5th edition, Ibid., p. 556.
15. Kandel, Eric R. et al. Principles of Neural Science, 5th edition Ibid., pp. 556-557.
16. Amthor, Frank. Neuroscience for Dummies, p. 93. Wiley & Sons, 2012.
17. Fuster, Joaquin, M. The Prefrontal Cortex, 4th edition, p. 260. Elsevier, 2008.
18. Fuster, Joaquin, M. The Prefrontal Cortex, 4th edition, Ibid., p. 260.
19. Kandel et al. Principles of Neural Science, 5th edition, Ibid., p. 413.
20. Amthor, Frank. Neuroscience for Dummies, Ibid., p. 166.
21. Gazzaniga, Michael S., Richard B. Irvy & George R. Mangun. Cognitive Neuroscience, 5th edition, Ibid., p. 578.
22. Solms, Mark & Oliver Turnball. The Brain and the Inner World, p. 51. Other Press, 2002.
23. Solms, Mark & Oliver Turnball. The Brain and the Inner World, Ibid., p. 52.
24. Pinker, Steven. How the Brain Works, Ibid., pp. 24-25.
25. Eagleman, David. Incognito, pp. 216-217. Pantheon Books, 2011.

[26] New England Institute of Ayurvedic Medicine, Manual I, p. 12. American Institute of Vedic Studies, 1996.

CHAPTER 10

[1] See: Fuster, Joaquin M. The Prefrontal Cortex, 4th edition. Elsevier, 2008.
[2] Fuster, Joaquin M. The Prefrontal Cortex, 4th edition, pp. 333-371. Elsevier, 2008.
[3] Fuster, Joaquin M. The Prefrontal Cortex, 4th edition, Ibid., p. 230.
[4] Fuster, Joaquin M. The Prefrontal Cortex, 4th edition, Ibid., p. 335.
[5] Fuster, Joaquin M. The Prefrontal Cortex, 4th edition, Ibid., pp. 336-338.
[6] Fuster, Joaquin M. The Prefrontal Cortex, 4th edition, Ibid., p. 258.
[7] Fuster, Joaquin M. The Prefrontal Cortex, 4th edition, Ibid., p. 341.
[8] Koziol, Leonard F. & Deborah Ely Budding. Subcortical Structures and Cognition, p. 11. Springer, 2010.
[9] Koziol, Leonard F. & Deborah Ely Budding. Subcortical Structures and Cognition, p. 50 citing Hazy, Frank, & O'Reilly (2007).
[10] Koziol, Leonard F. & Deborah Ely Budding. Subcortical Structures and Cognition, Ibid., p. 15 citing Richer & Chouinard (2003).
[11] Koziol, Leonard F. & Deborah Ely Budding. Subcortical Structures and Cognition, Ibid., p. 6 citing Adreasen & Pierson (2008).
[12] Koziol, Leonard F. & Deborah Ely Budding. Subcortical Structures and Cognition, Ibid., p. 48.
[13] Koziol, Leonard F. & Deborah Ely Budding. Subcortical Structures and Cognition, Ibid., pp. 58-60.
[14] Koziol, Leonard F. & Deborah Ely Budding. Subcortical Structures and Cognition, Ibid., p. 141.
[15] Koziol, Leonard F. & Deborah Ely Budding. Subcortical Structures and Cognition, Ibid., p. 126.
[16] Koziol, Leonard F. & Deborah Ely Budding. Subcortical Structures and Cognition, Ibid., p. 143.
[17] Kandel, Eric R. et al. Principles of Neural Science, 5th edition, pp. 335;423. McGraw Hill Medical, 2013.
[18] Dozier, Rush W. Jr. Why We Hate, p. 234. Contemporary Books, 2002.

CHAPTER 11

1. Winston, Robert. Human Instincts, p. 81. Bantam Books, 2003.
2. Cozolino, Louis. The Neuroscience of Human Relationships, p. 40 citing Schore (1994). W.W. Norton & Company, 2006.
3. Panksepp, Jaak. Affective Neuroscience, p. 250. Oxford University Press, 1998.
4. Panksepp, Jaak. Affective Neuroscience, Ibid., p. 248.
5. Panksepp, Jaak. Affective Neuroscience, Ibid., pp. 248-249.
6. Panksepp, Jaak. Affective Neuroscience, Ibid., p. 249.
7. Panksepp, Jaak. Affective Neuroscience, Ibid., 262.
8. Carter, Sue, James Harris, & Stephen W. Porges. *Neural and Evolutionary Perspectives on Empathy*, pp. 172-173. The Neuroscience of Empathy, edited by Jean Decety and William Ickes. A Bradford Book, MIT Press, 2011.
9. Cozolino, Louis. The Neuroscience of Human Relationships, Ibid., p. 55.
10. Amthor, Frank. Neuroscience for Dummies, pp. 174-175. Wiley & Sons, 2012.
11. Simpkins, C. Alexander & Annellen M. Simpkins. Neuroscience for Clinicians, p. 153. Springer, 2013.
12. Simpkins, C. Alexander &, Annellen M. Simpkins. Neuroscience for Clinicians, Ibid., p. 153.
13. Gazzaniga, Michael S., Richard B. Ivry & George R. Mangun. Cognitive Neuroscience, 3rd edition, pp. 92-93. W. W. Norton & Company, 2009.
14. Gazzanga, Michael S., Richard B. Ivry & George R. Mangun. Cognitive Neuroscience, 3rd edition, Ibid., p. 91.
15. See: Gazzaniga, Michael S., Richard B. Ivry & George R. Mangun. Cognitive Neuroscience, 3rd edition, pp. 92-93. W. W. Norton & Company, 2009.
16. See: Simpkins, Alexander C. & Annellen M. Simpkins. Neuroscience for Clinicians, pp. 151-160. Springer, 2013.
17. Simpkins, C. Alexander & Annellen M. Simpkins. Neuroscience for Clinicians, Ibid., p. 154.

18. Gazzaniga, Michael S., Richard B. Ivry & George R. Mangun. Cognitive Neuroscience, 3rd edition, Ibid., p. 94.
19. Gazzangia, Michael S., Richard B. Ivry & George R. Mangun. Cognitive Neuroscience, 3rd edition, Ibid., p. 99.
20. Kandel, Eric R. et al. Principles of Neural Science, 5th edition, p. 1233. McGraw Hill Medical, 2013.
21. Kandel, Eric R. et al. Principles of Neural Science, 5th edition, Ibid., p. 1213.
22. Amthor, Frank. Neuroscience for Dummies, Ibid., p. 333.
23. Amthor, Frank. Neuroscience for Dummies, Ibid., p. 333.
24. Kandel, Eric R. et al. Principles of Neural Science, 5th edition, Ibid., p. 1259.

CHAPTER 12

1. Cozolino, Louis. The Neuroscience of Psychotherapy, 2nd edition, p. 72. W.W. Norton & Company, 2010.
2. Cozolino, Louis. The Neuroscience of Human Relationships, p. 97. W.W. Norton & Company, 2006.
3. Cozolino, Louis. The Neuroscience of Human Relationships, Ibid., p. 98.
4. Cozolino, Louis. The Neuroscience of Psychotherapy, 2nd edition, Ibid., p. 183.
5. Eysenck, Michael W. & Mark T. Keane. Cognitive Psychology, A Student's Handbook, 6th edition, p. 297. Psychology Press, 2010.
6. Van der Kolk, Bessel, M.D. The Body Keeps the Score, p. 56. Viking, 2014.
7. Van der Kolk, Bessel, M.D. The Body Keeps the Score, Ibid., p. 56.
8. Fuster, Joaquin M. The Prefrontal Cortex, 4th edition, p. 207. Elsevier, 2008.
9. Cozolino, Louis. The Neuroscience of Psychotherapy, 2nd edition, Ibid., p. 19.
10. Rowe, John W., M.D. & Robert L. Kahn. Successful Aging, p. 135. Dell Trade Paperback, 1998.

11. Cozolino, Louis. The Neuroscience of Psychotherapy, 2nd edition, Ibid., p. 68.
12. Cozolino, Louis. The Neuroscience of Psychotherapy, 2nd edition, Ibid., p. 184.
13. Cozolino, Louis. The Neuroscience of Psychotherapy, 2nd edition, Ibid., pp. 184 citing Kehoe & Blass (1989).
14. Cozolino, Louis. The Neuroscience of Human Relationships, Ibid., p. 85 citing Cirulli et al. (2003).
15. Cozolino, Louis. The Neuroscience of Human Relationships, Ibid., p. 84 citing Meaney et al. (1989).
16. Cozolino, Louis. The Neuroscience of Human Relationships, Ibid., p. 114.
17. Cozolino, Louis. The Neuroscience of Human Relationships, Ibid., p. 250.
18. Cozolino, Louis. The Neuroscience of Psychotherapy, 2nd edition, Ibid., p. 156.
19. Eysenck, Michael W. & Mark T. Keane. Cognitive Psychology, A Student's Handbook, 6th edition, Ibid., p. 298.
20. Cozolino, Louis. The Neuroscience of Human Relationships, Ibid., p. 338.
21. Cozolino, Louis. The Neuroscience of Psychotherapy, 2nd edition, Ibid., p. 72.
22. Sousa, David A. How the Brain Learns, 4th edition, p. 26. Corwin, 2011.
23. Cozolino, Louis. Conference: How Therapy Changes the Brain - PESI Conference, August 6, 2010.
24. Sousa, David A. How the Brain Learns, 4th edition, Ibid., p. 27.
25. Eagleman, David. Incognito, p. 83. Pantheon Books, 2011.
26. Astington, Janet Wilde. The Child's Discovery of Mind, p. 44. Harvard University Press, 1993.
27. Astington, Janet Wilde. The Child's Discovery of Mind, Ibid., p. 44.
28. Baird, Jodie A. and Janet Wilde Astington. *The Development of the Intention Concept: From the Observable World to the Unobservable Mind*, p. 269. The New Unconscious, edited by Ran R. Hassin et al. Oxford University Press, 2005.
29. Astington, Janet Wilde. The Child's Discovery of Mind, Ibid., p.73.

30 Kellogg, Ronald T. The Making of the Mind, pp. 190-191. Prometheus Books, 2013.
31 Cozolino, Louis. The Neuroscience of Human Relationships, Ibid., p. 14.

CHAPTER 13

1 Cozolino, Louis. The Neuroscience of Psychotherapy, 2nd edition, p. 72. W.W. Norton & Company, 2010.
2 Solomon, Marion & Stan Tatkin. Love and War in Intimate Relationships, p. 88. W. W. Norton & Company, 2011.
3 See: Cozolino, Louis. The Neuroscience of Human Relationships. W.W. Norton & Company, 2006.
4 In this chapter, the terms "child," "infant," and "toddler" are used interchangeably.
5 Siegel, Daniel J. M.D. Pocket Guide to Interpersonal Neurobiology, pp. 20-2-3. W. W. Norton & Company, 2012.
6 Siegel, Daniel J. M.D. Pocket Guide to Interpersonal Neurobiology, Ibid., p. 21-4.
7 Siegel, Daniel J. M.D. Mindsight, p. 168. Bantam Books, 2010.
8 See: Siegel, Daniel J. M.D. Mindsight, pp. 168-171. Bantam Books, 2010.
9 Cozolino, Louis. The Neuroscience of Human Relationships, p. 143. W.W. Norton & Company, 2006.
10 Siegel, Daniel J. M.D. Mindsight, Ibid., p. 180.
11 Cozolino, Louis. The Neuroscience of Human Relationships, Ibid., p. 143.
12 Siegel, Daniel J. M.D. Mindsight, Ibid., p. 169.
13 Siegel, Daniel J. M.D. Mindsight, Ibid., pp. 168-169.
14 Cozolino, Louis. The Neuroscience of Psychotherapy, 2nd edition, Ibid., p. 191.
15 Cozolino, Louis. The Neuroscience of Psychotherapy, 2nd edition, Ibid., p. 181.
16 Cozolino, Louis. The Neuroscience of Psychotherapy, 2nd edition, Ibid., p. 204.
17 Siegel, Daniel J. M.D. Mindsight, Ibid., p. 174.

[18] Cozolino, Louis. The Neuroscience of Human Relationships, Ibid., p. 139.
[19] Ornish, Dean M.D. Love and Survival, p. 33. William Morrow, 1998.
[20] See: Siegel, Daniel J. M.D. Mindsight, pp. 171-186. Bantam Books, 2010. See also: Cozolino, Louis. The Neuroscience of Human Relationships, pp. 144-146. W.W. Norton & Company, 2006.
[21] Siegel, Daniel J. M.D. Mindsight, Ibid., p. 180.
[22] Cozolino, Louis. The Neuroscience of Human Relationships, Ibid., pp. 86.
[23] Cozolino, Louis. The Neuroscience of Human Relationships, Ibid., pp. 84; 86.
[24] Solms, Mark & Oliver Turnbull. The Brain and the Inner World, p. 30. Other Press, 2002.
[25] Cozolino, Louis. The Neuroscience of Psychotherapy, 2nd edition, Ibid., p. 193.
[26] Cozolino, Louis. The Neuroscience of Human Relationships, Ibid., p. 235.
[27] See Huber, Cheri. There's Nothing Wrong with You, revised edition. Keep It Simple Books, 2001.
[28] Leary, Mark. Understanding the Mysteries of Human Behavior, p. 49 of Course Guidebook. Great Courses, The Teaching Company, 2012.
[29] Porges, Stephen W. The Polyvagal Theory, cover. W. W. Norton & Company, 2011.
[30] Porges, Stephen W. The Polyvagal Theory, Ibid., p. 16.
[31] Porges, Stephen. The Polyvagal Theory, Ibid., p. 57.
[32] Simpkins, C. Alexander & Annellen M. Simpkins. The Dao of Neuroscience, p. 116. W. W. Norton & Company, 2010.
[33] Cozolino, Louis. The Neuroscience of Human Relationships, Ibid., p. 92.
[34] Solomon, Marion & Stan Tatkin, Love and War in Intimate Relationships, Ibid., pp. 227; 228.
[35] Cozolino, Louis. The Neuroscience of Psychotherapy, 2nd edition, Ibid., p. 334.
[36] See: Cozolino, Louis. The Neuroscience of Psychotherapy, 2nd edition, pp. 217-224. W.W. Norton & Company, 2010.
[37] Cozolino, Louis. The Neuroscience of Psychotherapy, 2nd edition, p. 223.

CHAPTER 14

1. Solomon, Marion & Stan Tatkin. Love and War in Intimate Relationships, p. 220. W.W. Norton & Company, 2011.
2. Gazzaniga, Michael S., Richard B. Ivry, & George R. Mangun. Cognitive Neuroscience, 3rd edition, G-4. W. W. Norton & Company, 2009.
3. Solms, Mark & Oliver Turnball. The Brain and the Inner World, p. 112. Other Press, 2002.
4. Siegel, Daniel M.D. Pocket Guide to Interpersonal Neurobiology, p. A1-26. W. W. Norton & Company, 2012.
5. Cacioppo, John T. et al. *The Affect System has Parallel and Integrative Processing Components: Form Follows Function*, p. 493. Foundations in Social Neuroscience, edited by John T. Cacioppo et al. A Bradford Book, MIT Press, 2002.
6. Brooks, David. The Social Animal, p. 12. Random House, 2011.
7. Modell, Arnold H. Imagination and the Meaningful Brain, pp. 152-153. A Bradford Book, MIT Press, 2006.
8. Shermer, Michael. The Believing Brain, pp. 104-105. Henry Holt and Company, 2011.
9. Solms, Mark & Oliver Turnbull. The Brain and the Inner World, Ibid., p. 278.
10. Leary, Mark. Understanding the Mysteries of Human Behavior, p. 43 of Course Guidebook. Great Courses, The Teaching Company, 2012.
11. Bechara, Antoine & Bar-on Reuven, *Substrates of Emotional and Social Intelligence*, pp. 14-15. Social Neuroscience, edited by John T. Cacioppo et al. A Bradford Book, MIT Press, 2006.
12. Kandel, Eric R. et al. Principles of Neural Science, 5th edition, p. 1080. McGraw Medical, 2013.
13. Bechara, Antoine & Bar-on Reuven, *Substrates of Emotional and Social Intelligence*, Ibid., p. 15.
14. Amthor, Frank. Neuroscience for Dummies, p. 324. Wiley & Sons, 2012.
15. Amthor, Frank. Neuroscience for Dummies, Ibid., p. 33.

CHAPTER 15

1. See: Panksepp, Jaak. Affective Neuroscience. Oxford University Press, 1998. See also: Solms, Mark & Oliver Turnbull. The Brain and the Inner World, pp. 114-137. Other Press, 2002.
2. Panksepp, Jaak. Affective Neuroscience, p. 188. Oxford University Press, 1998.
3. Panksepp, Jaak. Affective Neuroscience, Ibid., p. 207.
4. Panksepp, Jaak. Affective Neuroscience, Ibid., p. 215.
5. Gazzaniga, Michael S., Richard B. Ivry & George R. Mangun, Cognitive Neuroscience, 3rd edition, p. 372. W. W. Norton & Company, 2009.
6. Gazzaniga, Michael S., Richard B. Ivry & George R. Mangun, Cognitive Neuroscience, 3rd edition, Ibid., p. 372.
7. Amthor, Frank. Neursocience for Dummies, p. 323. Wiley & Sons, 2012.
8. Amthor, Frank. Neuroscience for Dummies, Ibid., p. 33.
9. Lawrence, Paul & Nitin Nohria. Driven, How Human Nature Shapes Our Choices, p. 131. Jossey-Bass, 2002.
10. Davidson, Richard J. & William Irwin. *The Functional Neuroanatomy of Emotion and Affective Style*, p. 481. Foundations in Social Neuroscience, edited by John Cacioppo et al. A Bradford Book, MIT Press, 2002.
11. Shapiro, Debbie. Your Body Speaks Your Mind, pp. 112- 119. The Crossing Press, 1997.
12. Cozolino, Louis. The Neuroscience of Psychotherapy, 2nd edition, p. 23. W.W. Norton & Company, 2010.
13. Eysenck, Michael W. & Mark T. Keane. Cognitive Psychology, A Student's Handbook, 6th edition, p. 572. Psychology Press, 2010.
14. Panksepp, Jaak. Affective Neuroscience, Ibid., p. 301.
15. Kandal, Eric R. et al. Principles in Neural Science, 5th edition, p. 1084. McGraw Medical, 2013.
16. Very little information is known about the connections of the human amygdala such that nonhuman primates provide the best estimate currently available.
17. Freese, Jennifer L. & David G. Amaral. *Neuroanatomy of the Primate Amygdala*, p. 15. The Human Amygdala, Ibid.

¹⁸ Freese, Jennifer L. & David G. Amaral. *Neuroanatomy of the Primate Amygdala*, Ibid., p. 16.
¹⁹ Freese, Jennifer L. & David G. Amaral. *Neuroanatomy of the Primate Amygdala*, Ibid., pp. 24-33.
²⁰ Freese, Jennifer L. & David G. Amaral. *Neuroanatomy of the Primate Amygdala*, Ibid., p. 34.
²¹ LeDoux, Joseph E. & Daniela Schiller. *The Human Amygdala, Insights from Other Animals*, Ibid., p. 52 citing Cardinal et al. (2002).
²² Gazzaniga, Michael S., Richard B. Ivry & George R. Mangun. Cognitive Neuroscience, 3rd edition, Ibid., p. 379.
²³ Buchanan, Tony W. et al. *The Human Amygdala in Social Function*, p. 289. The Human Amygdala, Ibid.
²⁴ Vuilleumier, Paul. *The Role of the Human Amygdala in Perception and Attention*, p. 22 citing Fecteau el al. (2007). The Human Amygdala, Ibid.
²⁵ Solomon, Marion & Stan Tatkin. Loe and War in Intimate Relationships, p. 220. W.W. Norton & Company, 2011.
²⁶ Gazzaniga, Michael S., Richard B. Ivry & George R. Mangun. Cognitive Neuroscience, 3rd edition, Ibid., pp. 623-625.
²⁷ Gazzaniga, Michael S., Richard B. Ivry & George R. Mangun. Cognitive Neuroscience, 3rd edition, Ibid., p. 625.
²⁸ Kandel, Eric R. et al. Principles of Neural Science, 5th edition, Ibid., p. 450.
²⁹ Mlodinow, Leonard. Subliminal, p. 206. Pantheon Books, 2012.
³⁰ Kandel, Eric R. et al. Principles of Neural Science, 5th edition, Ibid., p. 455.

CHAPTER 16

¹ Bor, Daniel. The Ravenous Brain, p. 84. Basic Books, 2012.
² LeDoux, Joseph. The Emotional Brain, p. 128. Simon and Schuster, 1996.
³ LeDoux, Joseph. The Emotional Brain, Ibid., p. 128.
⁴ Gazzaniga, Michael S., Richard B. Ivry & George R. Mangun. Cognitive Neuroscience, 3rd edition, pp. 371-372. W. W. Norton & Company, 2009.
⁵ LeDoux, Joseph. *Emotion: Clues from the Brain*, p. 392 citing Romanski & LeDoux (1993). Foundations in Social Neuroscience, edited by John T. Cacioppo et al. A Bradford Book, MIT Press, 2002.

[6] LeDoux, Joseph. Synaptic Self, p. 123. Penguin Books, 2002.
[7] Cozolino, Louis. The Neuroscience of Psychotherapy, 2nd edition, p. 312. W.W. Norton & Company, 2010
[8] Packer, Dominic J., Amanda Kesek & William A. Cunningham. *Self-Regulation and Evaluative Processing*, p. 151. Social Neuroscience, edited by Alexander Todorov et al., Oxford University Press, 2011.
[9] Packer, Dominic J., Amanda Kesek & William A. Cunningham. *Self-Regulation and Evaluative Processing.* Ibid., p. 154.
[10] LeDoux, Joseph. *Emotion: Clues from the Brain*, Ibid., p. 404.
[11] LeDoux, Joseph. The Emotional Brain, Ibid., p. 265 citing Amaral et al. (1992).
[12] Cozolino, Louis. The Neuroscience of Psychotherapy, 2nd edition, Ibid., p. 254.
[13] LeDoux, Joseph E. & Daniela Schiller. *The Human Amygdala, Insights from Other Animals*, p. 52. The Human Amygdala, edited by Paul J. Whalen & Elizabeth A. Phelps. The Guildford Press, 2009.
[14] Phelps, Elizabeth A. *The Human Amygdala and the Control of Fear*, p. 205. The Human Amygdala, Ibid.
[15] LeDoux, Joseph. The Synaptic Self, Ibid., p. 213.
[16] LeDoux, Joseph. *Emotion: Clues from the Brain*, Ibid., p. 399.
[17] LeDoux, Joseph. The Emotional Brain, Ibid., p. 291.
[18] Freese, Jennifer L. & David G. Amaral. *Neuroanatomy of the Primate Amydala*, p. 35. The Human Amygdala, Ibid.
[19] Amthor, Frank. Neuroscience for Dummies, pp. 152-183. Wiley & Sons, 2012.
[20] Amthor, Frank. Private Correspondence, February 18, 2015.
[21] Cozolino, Louis. The Neuroscience of Psychotherapy, 2nd edition, Ibid., p. 156.
[22] Taylor, Shelley E. et al. *Biobehavioral Responses to Stress in Females: Tend-and-Befriend, not Fight-or-Flight*, pp. 661-681. Foundations in Social Neuroscience, edited by John T. Cacioppo et al. A Bradford Book, MIT Press, 2002.

23 Taylor, Shelley E. et al., *Biobehavioral Responses to Stress in Females: Tend-and-Befriend, not Fight-or-Flight*, Ibid., p. 664.
24 Taylor, Shelley E. et al., *Biobehavioral Responses to Stress in Females: Tend-and-Befriend, not Fight-or-Flight*, Ibid., p. 665.
25 Taylor, Shelley E. et al. *Biobehavioral Responses to Stress in Females: Tend-and-Befriend, not Fight-or-Flight*, Ibid., p. 671.
26 Stress Effects and Stress Management, Psychological Harassment Information: www.psychologicalharassment.com/stress_and_stress_management.htm.
27 Van der Kolk, Bessel A. The Body Keeps the Score, p. 2. Viking, 2014.
28 Sapolsky, Robert. Biology and Human Behavior: The Neurological Origins of Individuality, p. 54 of Course Guidebook. Great Courses, The Teaching Company, 2005.
29 Van der Kolk, Bessel A. The Body Keeps the Score, Ibid., pp. 65; 82.
30 Brooks, David. The Social Animal, p. 12. Random House, 2011.

CHAPTER 17

1 Sousa, David A. How the Brain Learns, 4th edition, p. 48. Corwin, 2011.
2 Sousa, David A. How the Brain Learns, 4th edition, Ibid., p. 154.
3 Sousa, David A. How the Brain Learns, 4th edition, Ibid., p. 48.
4 NPR, All Things Considered. The 10 Questions Doctors Don't Want to Ask, quoting Dr. Jeff Brenner. March 3, 2015.
5 Remember, they are electromagnetic waves, pressure waves, and chemical compounds out of which the brain constructs sights, smells, etc.
6 McVeigh, Brian J. *The Self as Interiorized Social Relations*, p. 208. Reflections on the Dawn of Consciousness, edited by Marcel Kuijsten. Julian Jaynes Society, 2006.
7 Cozolino, Louis. The Neuroscience of Human Relationships, p. 132. W.W. Norton & Company, 2006.
8 Cozolino, Louis. The Neuroscience of Human Relationships, Ibid., p. 338.
9 Lieberman, Matthew D. Social, p. 185. Crown Publishers, 2013.
10 Lieberman, Matthew D. Social, Ibid., p. 186-187.

[11] Singer, Michael A. The Untethered Soul, pp. 100-130. Noetic Books, 2007. See also: the work of Pema Chodron and Cheri Huber.
[12] Simpkins, C. Alexander & Annellen M. Simpkins. Neuroscience for Clinicians, p. 40. Springer, 2013.
[13] Cozolino, Louis. The Neuroscience of Psychotherapy, 2nd edition, p. 135 citing Taylor & Brown (1988). W. W. Norton & Company, 2010.
[14] Burton, Robert A. M.D. On Being Certain, p. 45. St. Martin's Press, 2008.
[15] Cozolino, Louis. The Neuroscience of Psychotherapy, 2nd edition, Ibid., p. 314.
[16] Burton, Robert A. M.D. On Being Certain, Ibid., p. 160.
[17] Gazzaniga, Michael S., Richard B. Ivry & George R. Mangun. Cognitive Neuroscience, 3rd edition, p. 465. W. W. Norton & Company, 2009.
[18] Gazzaniga, Michael S., Richard B. Ivry & George R. Mangun Cognitive Neuroscience, 3rd edition, Ibid., p. 465.
[19] Kellogg, Ronald. The Making of the Mind, p. 108. Prometheus Books, 2013.
[20] Solms, Mark & Oliver Turnbull. The Brain and the Inner World, p. 77. Other Press, 2002.
[21] Kellogg, Ronald. The Making of the Mind, Ibid., p. 107.
[22] Kellogg, Ronald. The Making of the Mind, Ibid., p. 118.
[23] Cozolino, Louis. Neuroscience of Psychotherapy, 2nd edition, Ibid., p. 103.
[24] Gazzaniga, Michael S., Richard B. Ivry & George R. Mangun. Cognitive Neuroscience, 3rd edition, Ibid., p. 465.
[25] Kellogg, Ronald. The Making of the Mind, Ibid., p. 118.
[26] Kellogg, Ronald. The Making of the Mind, Ibid., pp. 240-241.
[27] Kellogg, Ronald. The Making of the Mind, Ibid., p. 185.
[28] Gazzaniga, Michael S., Richard B. Ivry & George R. Mangun. Cognitive Neuroscience, 3rd edition, Ibid., p. 465.
[29] Kellogg, Ronald. The Making of the Mind, Ibid., p. 110.
[30] See: Ruiz, Don Miguel. The Voice of Knowledge. Amber-Allen Publishing, 2004.
[31] Ruiz, Don Miguel. The Voice of Knowledge, p. 49. Amber-Allen Publishing, 2004.

32. Ruiz, Don Miguel. The Voice of Knowledge, Ibid., pp. 51; 53.
33. Amthor, Frank. Neuroscience for Dummies, p. 101. Wiley & Sons, 2012.
34. Siegel, Daniel J. M.D. Pocket Guide to Interpersonal Neurobiology, p. 14-5. W. W. Norton & Company, 2012.
35. Solms, Mark & Oliver Turnbull. The Brain and the Inner World, Ibid., p. 155.
36. Solms, Mark & Oliver Turnbull. The Brain and the Inner World, Ibid., p. 155.
37. Cozolino, Louis. Neuroscience of Psychotherapy, 2nd edition, Ibid., p. 313.
38. Sporns, Olaf. The Networks of the Brain, p. 206. MIT Press, 2011.
39. Cozolino, Louis. The Neuroscience of Human Relationships, p. 40 citing Schore (1994).

CHAPTER 18

1. Kandel, Eric R. In Search of Memory, p. 202. W. W. Norton & Company, 2006.
2. Solms, Mark & Oliver Turnball. The Brain and the Inner World, p. 147. Other Press, 2002.
3. Kandel, Eric R. In Search of Memory, Ibid., p. 202.
4. Sousa, David A. How the Brain Learns, 4th edition, p. 83. Corwin, 2011.
5. Dozier, Rush W. Jr. Why We Hate, p. 10. Contemporary Books, 2002.
6. Dozier, Rush W. Jr. Why We Hate, Ibid., p. 101.
7. Burton, Robert A. On Being Certain, p. 95. St. Martin's Griffin, 2008.
8. Burton, Robert A. On Being Certain, Ibid., p. 95.
9. Sousa, David A. How the Brain Learns, 4th edition, Ibid., p. 48.
10. See: Sousa, David A. How the Brain Learns, 4th edition, pp. 45-71. Corwin, 2011.
11. Sousa, David A. How the Brain Learns, 4th edition, Ibid., pp. 47-48.
12. Amthor, Frank. Neuroscience for Dummies, p. 272. Wiley & Sons, 2012.
13. Sousa, David A. How the Brain Learns, 4th edition, Ibid., p. 52.
14. Sousa, David A. How the Brain Learns, 4th edition, Ibid., p. 52.
15. Sousa, David A. How the Brain Learns, 4th edition, Ibid., p. 54.
16. Sousa, David A. How the Brain Learns, 4th edition, Ibid., p. 92.

17 Kandel, Eric R. In Search of Memory, Ibid., p. 210.
18 Sousa, David A. How the Brain Learns, 4th edition, Ibid., p. 126.
19 Sousa, David A. How the Brain Learns, 4th edition, Ibid., p. 127.
20 See: Sousa, David A. How the Brain Learns, 4th edition, p. 115. St. Martin's Griffin, 2008.
21 Sousa, David A. How the Brain Learns, 4th edition, Ibid., p. 104.
22 Sousa, David A. How the Brain Learns, 4th edition, Ibid., pp. 95-102.
23 Sousa, David A. How the Brain Learns, 4th edition, Ibid., p. 192.
24 Sousa, David A. How the Brain Learns, 4th edition, Ibid., p. 33 citing Bauerlein (2011).
25 Gazzaniga, Michael S., Richard B. Ivry, George R. Mangun. Cognitive Neuroscience, 3rd edition, p. 358. W. W. Norton & Company, 2009.
26 Amthor, Frank. Neuroscience for Dummies, Ibid., p. 269.
27 Amthor, Frank. Neuroscience for Dummies, Ibid., pp. 269-271.
28 Kandel et al., Principles in Neural Science, 5th edition, p. 1498. McGraw Medical, 2013.
29 Cozolino, Louis. The Neuroscience of Psychotherapy, 2nd edition, p. 57. W.W. Norton & Company, 2010.
30 Sousa, David A. How the Brain Learns, 4th edition, Ibid., p. 105.

CHAPTER 19

1 Gazzaniga, Michael S., Richard B. Ivry & George R. Mangun. Cognitive Neuroscience, 3rd edition, p. 360. W.W. Norton & Company, 2009.
2 Gazzaniga, Michael S., Richard B. Ivry & George R. Mangun. Cognitive Neuroscience, 3rd edition, Ibid., p. 325.
3 Tulving, Endel & Martin Lepage. *Where in the Brain is the Awareness of One's Past*, p. 209. Memory, Brain, and Belief, edited by Daniel L. Schacter & Elaine Scarry. Harvard University Press, 2000.
4 Sousa, David A. How the Brain Learns, 4th edition, p. 76. Corwin, 2011.
5 Gazzaniga, Michael S., Richard B. Ivry & George R. Mangun. Cognitive Neuroscience, 3rd edition, Ibid., p. 332.
6 Sousa, David A. How the Brain Learns, 4th edition, Ibid., p. 151.
7 Sousa, David A. How the Brain Learns, 4th edition, Ibid., p. 114.

[8] Lewis, Penelope A. The Secret World of Sleep, pp. 68-70. Palgrave Macmillan, 2013.
[9] Fuster, Joaquin M. The Prefrontal Cortex, 4th edition, pp. 338; 341. Elsevier, 2009.
[10] Siegel, Daniel J. M. D. Pocket Guide to Interpersonal Neurobiology, p. 30-2. W. W. Norton & Company, 2012.
[11] Sousa, David A. How the Brain Learns, 4th edition, Ibid., p. 114.
[12] Solms, Mark & Oliver Turnbull. The Brain and the Inner World, p. 140. Other Press, 2002.
[13] Solms, Mark & Oliver Turnbull. The Brain and the Inner World, Ibid., p. 139.
[14] Gazzaniga, Michael S., Richard B. Ivry & George R. Mangun. Cognitive Neuroscience, 3rd edition, Ibid., p. 322. Priming deals with the effects of prior exposure on the response to a stimulus; fear conditioning is an example of classical conditioning; habituation decreases a response to a stimulus while sensitization increases it.
[15] Gazzaniga, Michael S., Richard B. Ivry & George R. Mangun. Cognitive Neuroscience, 3rd edition, Ibid., p. 353.
[16] Gazzaniga, Michael S., Richard B. Ivry & George R. Mangun. Cognitive Neuroscience, 3rd edition, Ibid., p. 360.
[17] Gazzaniga, Michael S., Richard B. Ivry & George R. Mangun. Cognitive Neuroscience, 3rd edition, Ibid., p. 360.
[18] Gazzaniga, Michael S., Richard B. Ivry & George R. Mangun. Cognitive Neuroscience, 3rd edition, Ibid., p. 353.
[19] Gazzaniga, Michael S., Richard B. Ivry & George R. Mangun. Cognitive Neuroscience, 3rd edition, Ibid., p. 360.
[20] Sousa, David A. How the Brain Learns, 4th edition, Ibid., p. 88.
[21] Gazzaniga, Michael S., Richard B. Ivry & George R. Mangun. Cognitive Neuroscience, 3rd edition, Ibid., p. 559 citing Patricia Goldman-Rakic.
[22] Sousa, David A. How the Brain Learns, 4th edition, Ibid., p. 49.
[23] Gazzaniga, Michael S., Richard B. Ivry & George R. Mangun. Cognitive Neuroscience, 3rd edition, Ibid., p. 559.
[24] Amthor, Frank. Neuroscience for Dummies, Ibid., p. 275.

[25] Kandel, Eric R. In Search of Memory, p. 132. W.W. Norton & Company, 2006.
[26] Kandel, Eric R. In Search of Memory, Ibid., p. 132.
[27] Kandel, Eric R. In Search of Memory, Ibid., p. 132.
[28] Solomon, Marion & Stan Tatkin. Love and War in Intimate Relationships, p. 227. W.W. Norton & Company, 2011.
[29] Siegel, Daniel J. M.D. Mindsight, p. 155. Bantam Books, 2010.
[30] Solomon, Marion & Stan Tatkin. Love and War in Intimate Relationships, Ibid., p. 228.
[31] Siegel, Daniel J. M.D. Mindsight, Ibid., p. 165.
[32] Simpkins, C. Alexander & Annellen M. Simpkins, Neuroscience for Clinicians, p. 228. Springer, 2013.
[33] Simpkins, C. Alexander & Annellen M. Simpkins, Neuroscience for Clinicians, Ibid., p. 221.
[34] Simpkins, C. Alexander & Annellen M. Simpkins, Neuroscience for Clinicians, Ibid., p. 221.
[35] Eysenck, Michael W. & Mark T. Keane. Cognitive Psychology, A Student's Handbook, 6th edition, p. 255. Psychology Press, 2010.
[36] Eysenck, Michael W. & Mark T. Keane. Cognitive Psychology, A Student's Handbook, 6th edition, Ibid., p. 301.
[37] Eysenck, Michael W. & Mark T. Keane. Cognitive Psychology, A Student's Handbook, 6th edition, Ibid., p. 255 citing Tulving (2002).
[38] Eysenck, Michael W. & Mark T. Keane. Cognitive Psychology, A Student's Handbook, 6th edition, Ibid., p. 265 citing McCloskey & Gluckberg (1978).
[39] Eysenck, Michael W. & Mark T. Keane. Cognitive Psychology, A Student's Handbook, 6th edition, Ibid., p. 262 citing Schacter & Addis (2007).
[40] Sousa, David A. How the Brain Learns, 4th edition, Ibid., p. 88.
[41] Sousa, David A. How the Brain Learns, 4th edition, Ibid., p. 102.
[42] Sousa, David A. How the Brain Learns, 4th edition, Ibid., p. 102.
[43] Gazzaniga, Michael S., Richard B. Ivry & George R. Mangun. Cognitive Neuroscience, 3rd edition, Ibid., p. 356.
[44] Eichenbaum Howard & J. Alexander Bodkin, *Belief and Knowledge as Distinct Forms of Memory*, p. 177. Memory, Brain, and Belief, Ibid.

45 Eichenbaum Howard & J. Alexander Bodkin, *Belief and Knowledge as Distinct Forms of Memory*, Ibid., p. 197.
46 Eichenbaum Howard & J. Alexander Bodkin, *Belief and Knowledge as Distinct Forms of Memory*, Ibid., p. 202.
47 Eichenbaum Howard & J. Alexander Bodkin, *Belief and Knowledge as Distinct Forms of Memory*, Ibid., p. 202.
48 Eichenbaum Howard & J. Alexander Bodkin, *Belief and Knowledge as Distinct Forms of Memory*, Ibid., p. 202.
49 Schacter, Daniel L. *The Seven Sins of Memory*, p. 115. Foundations in Social Neuroscience, edited by John T. Capioppo et al. A Bradford Book, MIT Press, 2002.
50 Eysenck, Michael W. & Mark T. Keane. Cognitive Psychology, A Student's Handbook, 6th edition, Ibid., p. 244.
51 Siegel, Daniel L. Mindsight, Ibid., p. 148.
52 Sousa, David A. How the Brain Learns, 4th edition, Ibid., p. 123.
53 Schacter, Daniel L. *The Seven Sins of Memory*, Ibid., pp. 113-137.
54 Schacter, Daniel L. *The Seven Sins of Memory*, Ibid., p. 130.
55 Sousa, David A. How the Brain Learns, 4th edition, Ibid., p. 123.
56 Sousa, David A. How the Brain Learns, 4th edition, Ibid., p. 123.
57 Strauch, Barbara. *The Secret of the Grown-Up Brain*, New York Times, January 3, 2010.
58 See: Smith, Rodney. Lessons from the Dying, pp.75-90. Wisdom Publications, 1998.
59 Amthor, Frank. Neuroscience for Dummies, Ibid., p. 277.
60 Amthor, Frank. Neuroscience for Dummies, Ibid., p. 277.
61 Eysenck, Michael W. & Mark T. Keane. Cognitive Psychology, A Student's Handbook, 6th edition, Ibid., p. 258 citing Moscovitch et al. (2006).
62 Eysenck, Michael W. & Mark T. Keane. Cognitive Psychology, A Student's Handbook, 6th edition, Ibid., p. 262.
63 Solomon, Marion & Stan Tatkin. Love and War in Intimate Relationships, Ibid., p. 230.
64 Eichenbaum Howard & J. Alexander Bodkin, *Belief and Knowledge as Distinct Forms of Memory*, Ibid., p. 203.

CHAPTER 20

1. See: The New Unconscious, edited by Ran R. Hassin. Oxford University Press, 2005.
2. Gazzaniga, Michael S. *Brain and Conscious Experience*, p. 204. Foundations in Social Neuroscience, edited by John Cacioppo et al. A Bradford Book, MIT Press, 2002.
3. Mlodinow, Leonard. Subliminal, pp. 17-18. Pantheon Books, 2012.
4. Mlodinow, Leonard. Subliminal, Ibid., p. 33.
5. Solms, Mark & Oliver Turnbull. The Brain and the Inner World, p. 84 citing Bargh and Chartrand (1999). Other Press, 2002.
6. Mlodinow, Leonard. Subliminal, Ibid., p. 34.
7. Eagleman, David. Incognito, pp. 4-5. Pantheon Books, 2011.
8. Aronson, Elliot. The Social Animal, 11th edition, p. 153. Worth Publishers, 2012.
9. Siegel, Daniel J. M.D. Mindsight, p. 201. Bantam Books, 2010.
10. Siegel, Daniel J. M.D. Mindsight, Ibid., pp. 202-203.
11. Eagleman, David. Incognito, Ibid., p. 87.
12. Eagleman, David. Incognito, Ibid., p. 89.
13. Dozzier, Rush W. Why We Hate, pp. 40-43 citing Edward O. Wilson. Contemporary Books, 2002.
14. Eagleman, David. Incognito, Ibid., p. 56 citing Daniel L. Schacter.
15. Siegel, Daniel M.D. Mindsight, Ibid., p. 150.
16. Siegel, Daniel M.D. Mindsight, Ibid., p. 132.
17. Eysenck, Michael W. & Mark T. Keane. Cognitive Psychology, A Student's Handbook, 6th edition, p. 504 citing Shah & Oppenheimer (2008). Psychology Press, 2010.
18. Aronson, Elliot. The Social Animal, 11th edition, Ibid., p. 132 citing Kahneman & Tversky (1973).
19. Aronson, Elliot. The Social Animal, 11th edition, Ibid., p. 309.
20. Aronson, Elliot. The Social Animal, 11th edition, Ibid., p. 309.
21. See: Aronson, Elliot. The Social Animal, 11th edition, pp. 309-317. Worth Publishers, 2012.

22 Aronson, Elliot. The Social Animal, 11th edition, Ibid., pp. 308-309.
23 Ferguson, Melissa J. & Thomas C. Mann. *Effects of Evaluation: An Example of Robust "Social" Priming*, p. 36 citing Bruner (1957). Understanding Priming Effects in Social Psychology, edited by Daniel C. Molden. The Guilford Press, 2014.
24 Aronson, Elliot. The Social Animal, 11th edition, Ibid., p. 124 citing Bargh et al., (1996).
25 See: Shermer, Michael. The Believing Brain, pp. 274-276. Henry Holt and Company, 2011.
26 Aronson, Elliot. The Social Animal, 11th edition, Ibid., p. 119.
27 Cozolino, Louis. The Neuroscience of Psychotherapy, 2nd edition, pp. 136-137. W.W. Norton & Company, 2010.
28 Eysenck, Michael W. & Mark T. Keane. Cognitive Psychology, A Student's Handbook, 6th edition, Ibid., pp. 528 citing Galotti (2007); 522; 523 citing Galotti (2007).
29 Shermer, Michael. The Believing Brain, Ibid., pp. 259-261.
30 Eysenck, Michael W. & Mark T. Keane. Cognitive Psychology, A Student's Handbook, 6th edition, Ibid., p. 528.
31 Aronson, Elliot. The Social Animal, 11th edition, Ibid., p. 119.
32 Gazzaniga, Michael S. *Brain and Conscious Experience*, Ibid., p. 205.
33 Uleman, James S. et al. *Implicit Impressions*, pp. 372; 373. The New Unconscious, edited by Ran R. Hassin. Oxford University Press, 2005.
34 LeDoux, Joseph. The Synaptic Self, p. 23. Penguin Books, 2003.
35 See: Malle, Bertram F. *Folk Theory of Mind*. The New Unconscious, edited by Ran R. Hassin. Oxford University Press, 2005.
36 Choi, Y. Susan et al. *The Glimpsed World*, Ibid., p. 319.
37 Chartrand, Tanya L. et al. *Beyond the Perception-Behavior Link*, Ibid., p. 351.
38 Chartrand, Tanya L. et al. *Beyond the Perception-Behavior Link*, pp. 340; 343. The New Unconscious, Ibid.
39 Bargh, John A. *Bypassing the Will*, pp. 43-44. The New Unconscious, Ibid.

40. Glaser, Jack & John F. Kihlstrom. *Compensatory Automaticity*, p. 189. The New Unconscious, Ibid.
41. Roese, Neal J. et al. The *Mechanics of Imagination*, p. 138. The New Unconscious, Ibid.
42. Roese, Neal J. et al. The *Mechanics of Imagination*, Ibid., p. 161 citing Kahneman (1995).
43. Markman, Arthur B. & Dedre Gentner. *Nonintentional Similarity Processing*, Ibid., p. 130. Also Gazzaniga, Michael S. *Brain and Conscious Experience*, Ibid., p. 204.
44. Hassin, Ran R. *Nonconscious Control & Implicit Working Memory*, p. 204. The New Unconscious, Ibid.
45. Choi, Y. Susan, et al. *The Glimpsed World*, p. 323. The New Unconscious, Ibid.
46. Duhigg, Charles. The Power of Habit, p. xvi. Random House, 2012.
47. Duhigg, Charles. The Power of Habit, Ibid., p. 19.
48. NPR, Morning Edition, Today's Health. Interview with Wendy Wood, psychologist, University of Southern California. January, 2015.
49. McGonigal, Kelly. The Willpower Instinct, p. 56. Avery, 2012.
50. www.helpguide.org/harvard/addiction. Understanding Addiction. May 11, 2013.
51. Leonard F. Koziol & Deborah Ely Budding. Subcortical Structures and Cognition, Ibid., p. 35.
52. Fuster, Joaquin M. The Prefrontal Cortex, 4th edition, p. 345. Elsevier, 2009.
53. Eric Kandel, et al. Principles of Neural Science, 5th edition, p. 1110. McGraw Medical, 2013.
54. McGonigal, Kelly. The Willpower Instinct, p. 13 quoting Dr. Robert Sapolsky.
55. Gollwitzer, Peter M. et al. *The Control of the Unwanted*, p. 500. The New Unconscious, Ibid.
56. See: Duhigg, Charles. The Power of Habit. Random House, 2012.
57. Mlodinow, Leonard. Subliminal, Ibid., pp. 21; 20; 23; 27; 23; 24.
58. Mlodinow, Leonard. Subliminal, Ibid., p. 44.

59. Choi, Y. Susan, et al. *The Glimpsed World*, Ibid., p. 320 citing R.B. Zajonc (1980).
60. Kahneman, Daniel. Thinking, Fast and Slow, p. 348. Farrar, Straus and Giroux, 2011.
61. Mlodinow, Leonard. Subliminal, Ibid., p. 19.
62. Aronson, Elliot. The Social Animal, 11th edition, Ibid., p. 56.
63. Cozolino, Louis. The Neuroscience of Human Relationships, p. 185. W.W. Norton & Company, 2006.
64. Markman, Arthur B. & Dedre Gentner. *Nonintentional Similarity Processing*, p. 131. The New Unconscious, Ibid.
65. Mlodinow, Leonard. Subliminal, Ibid., p. 122.
66. Brooks, David. Social Animal, pp. 8-9. Random House, 2011.
67. Todorov, Alexander. *Evaluating Faces on Social Dimensions*, p. 54 citing Todorov et al. (2009). Social Neuroscience, edited by Alexander Todorov et al. Oxford University Press, 2011.
68. Everyday Health.com/ how-our-brain-decides-who-we-can-trust. August 9, 2014 citing an article in the Journal of Neuroscience.
69. Mlodinow, Leonard. Subliminal, Ibid., pp. 119; 123; 121.

CHAPTER 21

1. Amthor, Frank. Neuroscience for Dummies, p. 336. Wiley & Sons, 2012.
2. Eysenck, Michael W. & Mark T. Keane. Cognitive Psychology, A Student's Handbook, 6th edition, p. 614. Psychology Press, 2010.
3. Eysenck, Michael W. & Mark T. Keane. Cognitive Psychology, A Student's Handbook, 6th edition, Ibid., p. 624.
4. Amthor, Frank. Neuroscience for Dummies, Ibid., p. 336.
5. Bor, Daniel. The Ravenous Brain, p. xiv. Basic Books, 2012.
6. LeDoux, Joseph. The Emotional Brain, pp. 281-282. Simon & Schuster, 1996.
7. Bargh, John A. *Bypassing the Will*, p. 53. The New Unconscious, edited by Ran R. Hassin et al. Oxford University Press, 2005.
8. LeDoux, Joseph. The Emotional Brain, Ibid., p. 33.

[9] Siegel, Daniel J. M.D. Pocket Guide for Interpersonal Neurobiology, p. 24-4. W. W. Norton & Company, 2012.
[10] Amthor, Frank. Neuroscience for Dummies, Ibid., p. 199.
[11] Amthor, Frank. Neuroscience for Dummies, Ibid., p. 200.
[12] Sousa, David A. How the Brain Learns, 4th edition, p. 112. Corwin, 2011.
[13] Sousa, David A. How the Brain Learns, 4th edition, Ibid., p. 113.
[14] Amthor, Frank. Neuroscience for Dummies, Ibid., pp. 207-208.
[15] Amthor, Frank. Neuroscience for Dummies, Ibid., p. 218.
[16] Smith, Rodney. Lessons from the Dying, p. 84. Wisdom Publications, 1998.
[17] See: Smith, Rodney. Lessons from the Dying, p. 82-85. Wisdom Publications, 1998.
[18] Amthor, Frank. Neuroscience for Dummies, Ibid., p. 201.
[19] Amthor, Frank. Neuroscience for Dummies, Ibid., p. 214.
[20] Fuster, Joaquin M. The Prefrontal Cortex, 4th edition, p. 134. Elsevier, 2009.
[21] Bergland, Christopher. *The Neuroscience of Imagination*. Psychologytoday.com/blog/the-athletes-way/201202. December 17, 2012.
[22] Lehrer, Jonah. Imagination, pp. 104-107. Houghton Mifflin Harcourt, 2012.
[23] Jung, Rex. *Creativity and the Everyday Brain*, p. 8 of transcript. On Being with Krista Tippett, May 2, 2013.
[24] Jung, Rex. *Creativity and the Everyday Brain*, Ibid., p. 6 of transcript.
[25] Kounios, John & Mark Beeman, *The Aha! Moment*, p. 210. Current Directions in Psychological Science 2009 18: 210.
[26] John Kounios, private correspondence, March 31, 2012.
[27] Hotz, Robert Lee. *A Wandering Mind Heads Straight Toward Insight*, p. 3. The Wall Street Journal, March 22, 2012.
[28] Kounios, John & Mark Beeman, *The Aha! Moment*, Ibid., p. 212.
[29] Kandel et al. Principles of Neural Science, 5th edition, pp. 384-385. McGraw Medical, 2013.
[30] Wegner, Daniel M. *Who is the Controller of Controlled Processes?* p. 23 citing Wegner (2002). The New Unconscious, Ibid.

31 Wegner, Daniel M. *Who is the Controller of Controlled Processes?* Ibid., p. 30.
32 Wegner, Daniel M. *Who is the Controller of Controlled Processes?* Ibid., p. 30.
33 Bor, Daniel. The Ravenous Brain, Ibid., p. 103.
34 Eysenck, Michael W. & Mark T. Keane. Cognitive Psychology, A Student's Handbook, 6th edition, Ibid., p. 611.
35 Cozolino, Louis. The Neuroscience of Psychotherapy, 2nd edition, p. 313. W.W. Norton & Company, 2012.
36 Eysenck, Michael W. & Mark T. Keane. Cognitive Psychology, A Student's Handbook, 6th edition, Ibid., p. 610.
37 Wegner, Daniel M. *Who is the Controller of Controlled Processes?* Ibid., p. 30.
38 Eysenck, Michael W. & Mark T. Keane. Cognitive Psychology, 6th edition, A Student's Handbook, Ibid., p. 611 quoting John-Dylan Hayes (2008).
39 Eagleman, David. Incognito, pp. 176-177. Pantheon, 2011.
40 Eagleman, David. Incognito, Ibid., p. 177.
41 Fuster, Joaquin M. The Prefrontal Cortex, 4th edition, Ibid., pp. 340-344.
42 Eysenck, Michael W. & Mark T. Keane. Cognitive Psychology, A Student's Handbook, 6th edition, Ibid., p. 614.
43 Eysenck, Michael W. & Mark T. Keane. Cognitive Psychology, A Student's Handbook, 6th edition, Ibid., p. 614.
44 Siegel, Daniel J. M.D. Pocket Guide for Interpersonal Neurobiology, Ibid., pp. 17-5; 17-4; 17-5.
45 Siegel, Daniel J. M.D. Mindsight, p. 86. Bantam Books, 2010.
46 Davidson, David. Investigating Healthy Minds, p. 4 of transcript. On Being with Krista Tippett, June 14, 2012.
47 Smith, Jonathan C. *Alterations in Brain and Immune Function Produced by Mindfulness Meditation: Three Caveats,* p. 149. Psychosomatic Medicine, Volume 66, January 1, 2004.
48 This is the title of a book by Allan Combs (Paragon House, 2009) which examines Ken Wilber's and Allan Combs' work on consciousness.

49 Hampton, Simon. Essential Evolutionary Psychology, pp. 77-78. Sage, 2010.
50 Hampton, Simon. Essential Evolutionary Psychology, Ibid., p. 78
51 Bor, Daniel. The Ravenous Brain, Ibid., p. 107.
52 New England Institute of Ayurvedic Medicine, Manual 1. American Institute of Vedic Studies, 1996.
53 New England Institute of Ayurvedic Medicine, Manual 1. American Institute of Vedic Studies, 1996.
54 Dozier, Rush W. Why We Hate, pp. 228-233. Contemporary Books, 2002.

CHAPTER 22

1 Adolfs, Ralph. *What is Special about Social Cognition?* p. 270. Social Neuroscience -- People Thinking about Thinking People, edited by John T. Capioppo et al. A Bradford Book, MIT Press, 2006.
2 Adolfs, Ralph. *What is Special about Social Cognition?* Ibid., p. 271.
3 Amthor, Frank. Neuroscience for Dummies, p. 255. Wiley & Sons, 2012.
4 Shermer, Michael. The Believing Brain, p. 131. Times Books, 2011.
5 Banaji, Mahzarin R. Foreword, p. viii. Social Neuroscience -- People Thinking about Thinking People, Ibid.
6 Siegel, Daniel J. M.D. Pocket Guide to Interpersonal Neurobiology, p. AI-75. W. W. Norton & Company, 2012.
7 Jenkins, Adrianna C. & Jason P. Mitchell. *How Has Cognitive Neuroscience Contributed to Social Psychological Theory?* p. 4. Social Neuroscience, edited by Alexander Todorov et al. Oxford University Press, 2011.
8 Mitchell, Jason P. et al. *Thinking about Others: The Neural Substrates of Social Cognition*, p. 63 citing Adolfs (1999). Social Neuroscience -- People Thinking about Thinking People, Ibid.
9 Lieberman, Matthew D. Social, p. 8. Crown Publications, 2013.
10 Lieberman, Matthew D. Social, Ibid., p. 26.
11 Lieberman, Matthew D. Social, Ibid., p. 19.
12 Lieberman, Matthew D. Social, Ibid., p. 215.

13. Lieberman, Matthew D. Social, Ibid., p. 183.
14. Lieberman, Matthew D. Social, Ibid., p. 185.
15. Lieberman, Matthew D. Social, Ibid., pp. 186-187.
16. Gobbini, Maria Ida. *Distributed Process for Retrieval of Person Knowledge*, p. 41. Social Neuroscience, Ibid.
17. Mitchell, Jason P. et al. *Thinking about Others: The Neural Substrates of Social Cognition*, Ibid., p. 63 citing Adolfs (1999).
18. Lieberman, Matthew D. Social, Ibid., p. 8.
19. Jenkins, Adrianna C. & Jason P. Mitchell. *How has Cognitive Neuroscience Contributed to Social Psychology Theory?* Ibid., p. 10.
20. Jenkins, Adrianna C. & Jason P. Mitchell. *How has Cognitive Neuroscience Contributed to Social Psychology Theory?* Ibid., p. 10.
21. Adolfs, Ralph. *What is Special about Social Cognition?* Ibid. p. 281.

CHAPTER 23

1. See: Sapolsky, Robert. The Biology of Human Behavior, pp. 58-61 of Course Guidebook. Great Courses, The Teaching Company, 2005.
2. Sapolsky, Robert. The Biology of Human Behavior, pp. 59 of Course Guidebook. Great Courses, The Teaching Company, 2005.
3. Haxby, James V. *Social Neuroscience and the Representation of Others: Commentary*, p. 77 citing Brothers (1990), Dunbar (1998). Social Neuroscience, edited by Alexander Todorov et al. Oxford University Press, 2011.
4. Lieberman, Matthew D. Social, p. 32. Crown Books, 2013.
5. Hampton, Simon. Essential Evolutionary Psychology, Ibid., pp. 156-157.
6. Stone, Valerie E. *Theory of Mind and the Evolution of Social Intelligence*, p. 103. Social Neuroscience -- People Thinking about Thinking People, edited by John T. Capioppo et al. A Bradford Book, MIT Press, 2006.
7. Stone, Valerie E. *Theory of Mind and the Evolution of Social Intelligence*, Ibid., p. 103.
8. Kellogg, Ronald T. The Making of the Mind, p. 59. Prometheus Books, 2013.

9   Kellogg, Ronald T. The Making of the Mind, Ibid., p. 58.
10 Renfrew, Colin. Prehistory, The Making of the Human Mind, p. 82. Modern Library, 2007.
11 Kellogg, Ronald T. The Making of the Mind, Ibid., p. 57.
12 Watson, Mallcolm W. Theories of Human Development, p. 128 of Course Guidebook. Great Courses, The Teaching Company, 2002.
13 Brooks, David. The Social Animal, p. 149. Random House, 2011.
14 Greer, Scott. *A Knowing Noos and a Slippery Psyche*, p. 234. Reflections on the Dawn of Consciousness, edited by Marcel Kuijsten. Julian Jaynes Society, 2006.
15 Kuijsten, Marcel. *Consciousness, Hallucinations, and the Bicameral Mind*, p. 116. Reflections on the Dawn of Consciousness, Ibid.
16 Greer, Scott. *A Knowing Noos and a Slippery Psyche*, Ibid., p. 234.
17 McVeigh, Brian J. *The Self as Interiorized Social Relations*, Ibid., p. 214.
18 Greer, Scott. *A Knowing Noos and a Slippery Psyche*, Ibid., p. 244.
19 McVeigh, Brian J. *The Self as Interiorized Social Relations*, Ibid., p. 213.
20 McVeigh, Brian J. *The Self as Interiorized Social Relations*, Ibid., p. 214.
21 McVeigh, Brian J. *The Self as Interiorized Social Relations*, Ibid., p. 214.
22 Pagel, Mark. Wired for Culture, p. 26. W. W. Norton & Company, 2012.
23 Pagel, Mark. Wired for Culture, Ibid., pp. 7; 5.
24 Panksepp, Jaak. Affective Neuroscience, Ibid., p. 162.
25 Kellogg, Ronald T. The Making of the Mind, Ibid., p. 65.
26 McVeigh, Brian J. *The Self as Interiorized Social Relations*, Ibid., p. 211.
27 McVeigh, Brian J. *The Self as Interiorized Social Relations*, Ibid., p. 206.
28 Lieberman, Matthew. Social, pp. 190; 227. Crown Publishers, 2013.
29 McVeigh, Brian J. *The Self as Interiorized Social Relations*, Ibid., p. 204.
30 McVeigh, Brian J. *The Self as Interiorized Social Relations*, Ibid., p. 214.
31 Pagel, Mark. Wired for Culture, Ibid., p. 53.
32 See: Pagel, Mark. Wired for Culture, pp. 69-98. W. W. Norton & Company, 2012.
33 Hampton, Simon. Essential Evolutionary Psychology, p. 92 citing Tajfel (1970). Sage, 2010.
34 Lieberman, Matthew. Social, Ibid., p. 228.
35 Lieberman, Matthew. Social, Ibid., p. 229.

[36] Aronson, Elliot. The Social Animal, 11th edition, p. 357. Worth Publishers, 2012.
[37] Heatherton, Todd F. *Neuroscience of Self and Self-Regulation*, p.1. NIH Public Access Author Manuscript. www.ncbi.nim.nih.gobe/pmc/articles/PMC3056504.
[38] Heatherton, Todd. F. *Neuroscience of Self and Self-Regulation*, Ibid., p. 1
[39] Lieberman, Matthew. Social, Ibid., pp. 226,232.
[40] Norretranders, Tor. *Altruism*, p. 212. This Idea Must Die, Harper Perennial. 2015.
[41] Sapolsky, Robert. Biology and Human Behavior, citing a Prisoner's Dilemma study. Great Courses, The Teaching Company, 2005.
[42] Decety, Jean & Claus Lamm. *Empathy versus Personal Distress*, p. 208. The Social Neuroscience of Empathy, edited by Jean Decety & William Ickes. The Bradford Book, MIT Press, 2011.
[43] Decety, Jean & Claus Lamm. *Empathy versus Personal Distress*, Ibid., p. 208 citing Lanzetta & Englis (1989). Lieberman, Matthew. Social, Ibid., p. 231 citing Cooley & Mead.
[44] Lieberman, Matthew. Social, Ibid., p. 231 citing Cooley & Mead.
[45] Chartrand, Tanya L. et al. *Beyond the Perception-Behavior Link*, p. 343. The New Unconscious, edited by Ran R. Hassin et al. Oxford University Press, 2005
[46] Lieberman, Matthew. Social, Ibid., p. 91.
[47] Lieberman, Matthew. Social, Ibid., p. 95.
[48] Riskin, Dan. Mother Nature is Trying to Kill You, pp. 18-20. Touchstone, 2014.
[49] Brown, Brene. Daring Greatly, p. 232. Gotham Books, 2012.
[50] Waytz, Adam. *Humans Are by Nature Social Animals*, pp.217-218. This Idea Must Die, Ibid.

CHAPTER 24

[1] Ambady, Nalini et al. *Race and Emotion: Insights from Social Neuroscience Perspective*, p. 211 citing Ekman (1992). Social Neuroscience – People Thinking about Thinking People, edited by John T. Cacioppo et al. The Bradford Book, MIT Press, 2006.

2. Ambady, Nalini et al. *Race and Emotion: Insights from Social Neuroscience Perspective*, Ibid., p. 210 citing Eibl-Eibesfeldt (1970).
3. Cozolino, Louis. The Neuroscience of Human Relationships, p. 171 citing Lu et al. (1991). W. W. Norton & Company, 2006.
4. Gobbini, Maria Ida. *Distributed Process for Retrieval of Person Knowledge*, p. 41. Social Neuroscience, edited by Alexander Todorov et al. Oxford University Press, 2011.
5. Gobbini, Maria Ida. *Distributed Process for Retrieval of Person Knowledge*, Ibid., p. 43.
6. Gazzaniga, Michael S., Richard B. Ivry & George R. Mangun. Cognitive Neuroscience, 3rd edition, p. 459. W. W. Norton & Company, 2009.
7. Whalen, Paul J. et al. *Human Amygdala Responses to Facial Expressions of Emotions*, p. 271. The Human Amygdala, edited by Paul J. Whalen & Elizabeth A. Phelps. Guildford Press, 2009.
8. Cozolino, Louis. The Neuroscience of Human Relationships, Ibid., p. 182.
9. Aronson, Elliot. The Social Animal, 11th edition, p. 309. Worth Publishers, 2012.
10. Brooks, David. The Social Animal, 11th edition, pp. 8-9 citing Todorov & Willis. Random House, 2011.
11. Ito, Tiffany A. et al. *The Social Neuroscience of Stereotyping and Prejudice: Using Event-Related Brain Potentials to Study Social Perception*, pp. 195-196. Social Neuroscience – People Thinking about Thinking People, Ibid.
12. Ito, Tiffany A. *Perceiving Social Category Information from Faces: Using ERPs to Study Person Perception*, p. 97. Social Neuroscience, edited by Alexander Todorov et al., Ibid.
13. Heatherton, Todd. F. *Neuroscience of Self and Self-Regulation*, p. 5 citing Jenkins et al. (2008). NIH Public Access Author Manuscript. www.ncbi.nlm.nih.gov/pmc/articles/PMC3056504.
14. Amodio, David M. *Self-Regulation in Intergroup Relations: A Social Neuroscience Framework*, p. 106. Social Neuroscience, edited by Alexander Todorov et al., Ibid.

15. Ambady, Nalini et al. *Race and Emotion: Insights from Social Neuroscience Perspective*, Ibid., p. 136.
16. See: Eisenberger, Naomi I. *Social Pain: Experiential, Neurocognitive, and Genetic Correlates*, p. 229-248. Social Neuroscience, edited by Alexander Todorov et al. Oxford University Press, 2011.
17. Lieberman, Matthew D. & Naomi I. Eisenberger. *A Pain by Any Other Name*, p. 167. Social Neuroscience – People Thinking about Thinking People, Ibid.
18. Lieberman, Matthew D. & Naomi I. Eisenberger. *A Pain by Any Other Name*, Ibid, p. 169.
19. Lieberman, Matthew D. & Naomi I. Eisenberger. *A Pain by Any Other Name*, Ibid, p. 175.
20. Eisenberger, Naomi I. *Social Pain: Experiential, Neurocognitive, and Genetic Correlates*, p. 239. Social Neuroscience, edited by Alexander Todorov et al., Ibid.
21. Watson, Mallcolm W. Theories of Human Development, p. 129 of Course Guidebook. Great Courses, The Teaching Company, 2002.
22. Watson, Mallcolm W. Theories of Human Development, Ibid., p. 129 of Course Guidebook.
23. Sousa, David A. How the Brain Learns, 4th edition, p. 189. Corwin, 2011.
24. Sousa, David A. How the Brain Learns, 4th edition, Ibid., pp. 190; 194-195.
25. Eysenck, Michael W. & Mark T. Keane. Cognitive Psychology, A Student's Handbook, 6th edition, p. 418. Psychology Press, 2010.
26. Eysenck, Michael W. & Mark T. Keane. Cognitive Psychology, A Student's Handbook, 6th edition, Ibid., p. 421.
27. Eysenck, Michael W. & Mark T. Keane. Cognitive Psychology, A Student's Handbook, 6th edition, Ibid., pp. 424-425.
28. Cozolino, Louis. The Neuroscience of Psychotherapy, 2nd edition, pp. 170-171. W. W. Norton & Company, 2010.
29. Cozolino, Louis. The Neuroscience of Psychotherapy, 2nd edition, Ibid., pp. 170-171.
30. Aronson, Elliot. The Social Animal, 11th edition, Ibid., p. 31.

31 McGonigal, Kelly. The Willpower Instinct, p. 193. Avery, 2012.
32 McGonigal, Kelly. The Willpower Instinct, Ibid., p. 193.
33 Astington, Janet Wilde. The Developing Child, p.135. Harvard University Press, 1993.
34 Choi, Y. Susan et al. *The Glimpsed World: Unintended Communication and Unintended Perception*, p. 316. The New Unconscious, edited by Ran R. Hassin. Oxford University Press, 2005.
35 Choi, Y. Susan et al. *The Glimpsed World: Unintended Communication and Unintended Perception*, Ibid., pp. 316-317.
36 Cozolino, Louis. The Neuroscience of Psychotherapy, 2nd edition, Ibid., p. 136.
37 Lakens, Daniel. *Grounding Social Embodiment*, p. 179. Understanding Priming Effects in Social Psychology, edited by Daniel C. Molden. The Guilford Press, 2014.
38 Aronson, Elliot. The Social Animal, 11th edition, Ibid., p. 261.
39 Rosenbaum, Thane. Payback, p. 212. University of Chicago Press, 2013.
40 Rosenbaum, Thane. Payback, Ibid., p. 27.
41 Rosenbaum, Thane. Payback, pp. 99; 91-92.
42 Rosenbaum, Thane. Payback, Ibid., pp. 93; 2.

CHAPTER 25

1 Epley, Nicholas. Mindwise, p. xi. Alfred A. Knopf, 2014.
2 Harris, Lasana T. & Susan T. Fiske. *Perceiving Humanity or Not*, p. 124. Social Neuroscience, edited by Alexander Todorov. Oxford University Press, 2011.
3 Samson, Dana & Caroline Michel. *Theory of Mind: Insights from Patients with Acquired Brain Damage*, p. 164. Understanding Other Minds, edited by Simon Baron-Cohen et al. Oxford University Press, 2013.
4 Gazzaniga, Michael S., Richard B. Ivry & George R. Mangun. Cognitive Neuroscience, 3rd edition, p. 608. W. W. Norton & Company, 2009.
5 Gazzaniga, Michael S., Richard B. Ivry & George R. Mangun. Cognitive Neuroscience, 3rd edition, Ibid., p. 609.

6. Gobbini, Maria Ida. *Distributed Process for Retrieval of Person Knowledge*, p. 45 citing Allison et al. (2000). Social Neuroscience, Ibid.
7. Gobbini, Maria Ida. *Distributed Process for Retrieval of Person Knowledge*, Ibid., p.45.
8. Hamilton, Antonia & Lauren Marsh. *Two Systems for Action Comprehension*, pp. 382-383 citing Saxe & Powell (2006) and Fletcher et al. (1995). Understanding Other Minds, Ibid.
9. Shermer, Michael. The Believing Brain, pp. 131-132. Times Books, 2011.
10. Carruthers, Peter. *Mindreading the Self*, pp. 467-468. Understanding Other Minds, Ibid.
11. Carruthers, Peter. *Mindreading the Self*, Ibid., pp. 467-468.
12. Samson, Dana & Caroline Michel. *Theory of Mind: Insights from Patients with Acquired Brain Damage*, Ibid., pp. 165-173.
13. Lieberman, Matthew D. Social, p. 118. Crown Publishers, 2013.
14. Harris, Lasana T. & Susan T. Fiske. *Perceiving Humanity or Not*, Ibid., p. 124.
15. Pyers, Jennie & Peter A. de Villiers. *Theory of Mind in Deaf Children*, p. 358. Understanding Other Minds, Ibid.
16. Pyers, Jennie & Peter A. de Villiers. *Theory of Mind in Deaf Children*, Ibid., pp. 346; 349.
17. Astington, Janet Wilde. The Child's Discovery of the Mind, p. 59. Harvard University Press, 1993.
18. Stone, Valerie E. *Theory of Mind and the Evolution of Social Intelligence*, p. 107, citing Suddendorf (1999). Social Neuroscience -- People Thinking about Thinking People, edited by John T Cacioppo et al. A Bradford Book, MIT Press. 2006.
19. Stone, Valerie E. *Theory of Mind and the Evolution of Social Intelligence*, Ibid., p. 107 citing Woodward (1999) and Csibra et al. (2003), respectively.
20. Kellogg, Ronald T. The Making of Mind, p. 68. Prometheus Books, 2013.
21. Kellogg, Ronald T. The Making of Mind, Ibid., p. 68 quoting Michael Tomasello.

22 Stone, Valerie E. *Theory of Mind and the Evolution of Social Intelligence*, Ibid., p. 109.
23 Stone, Valerie E. *Theory of Mind and the Evolution of Social Intelligence*, Ibid., p. 109.
24 Stone, Valerie E. *Theory of Mind and the Evolution of Social Intelligence*, Ibid., p. 112.
25 Astington, Janet Wilde. The Child's Discovery of the Mind, Ibid., p. 160.
26 Astington, Janet Wilde. The Child's Discovery of the Mind, Ibid., p. 82.
27 Astington, Janet Wilde. The Child's Discovery of the Mind, Ibid., p. 83.
28 Astington, Janet Wilde. The Child's Discovery of the Mind, Ibid., p. 121.
29 Wellman, Henry M. & Candida C. Peterson. *Theory of Mind, Development, and Deafness*, p. 54. Understanding Other Minds, Ibid.
30 Astington, Janet Wilde. The Child's Discovery of the Mind, Ibid., p. 112.
31 Apperly, Ian. Can Theory of Mind Grow Up? P. 85. Understanding Other Minds, Ibid.
32 Miller, Scott A. Theory of Mind beyond the Preschool Years, p. 79 citing Bosacki and Astington (1999). Psychology Press, 2012.
33 Perry, Anat & Simone Shamay-Tsoory. *Understanding Emotional and Cognitive Empathy*, p. 179. Understanding Other Minds, Ibid.
34 Lieberman, Matthew D. Social, Ibid., p. 133.
35 Fuster, Joaquin M. The Prefrontal Cortex, 4th edition, p. 343. Elsevier, 2008.
36 Lieberman, Matthew D. Social, Ibid., p. 135.
37 Gazzaniga, Michael S., Richard B. Ivry & George R. Mangun. Cognitive Neuroscience, 3rd edition, Ibid., p. 283.
38 Lieberman, Matthew D. Social, Ibid., p. 135.
39 Simpkins, C. Alexander & Annellen M. Simpkins. The Dao of Neuroscience, p. 219. W.W. Norton & Company, 2010.
40 Keysers, Christian, Marc Thioux & Valeria Gazzola. *Mirror Neuron System and Social Cognition*, p. 251. Understanding Other Minds, Ibid.
41 Cozolino, Louis. The Neuroscience of Psychotherapy, 2nd edition, p. 232. W.W. Norton & Company, 2010.
42 Lieberman, Matthew D. Social, Ibid., p. 140.

43 Simpkins, C. Alexander & Annellen M. Simpkins. The Dao of Neuroscience, Ibid., p. 219.
44 Ramachandran, V.S. The Tell-Tale Brain, pp. 128-130. W.W. Norton & Company, 2011.
45 Iacoboni, Marco. Mirroring People, p. 76. Picador, 2008.
46 Keysers, Christian, Marc Thioux & Valeria Gazzola. *Mirror Neuron System and Social Cognition*, Ibid. p. 249 citing Thioux et al. (2008).
47 Samson, Dana & Caroline Michel. *Theory of Mind: Insights from Patients with Acquired Brain Damage*, Ibid., p. 170.
48 Whiten, Andrew. *Culture and the Evolution of Interconnected Minds*, p. 437 citing Stone & Davies (1996). Understanding Other Minds, Ibid.
49 See: Understanding Other Minds, edited by Simon Baron-Cohen et al. Oxford University Press, 2013.
50 Zaki, Jamil & Kevin Ochsner. *You, Me, and My Brain*, p. 19. Social Neuroscience, Ibid.
51 Lieberman, Matthew D. Social, Ibid., p. 144.
52 Lieberman, Matthew D. Social, Ibid., p. 132.
53 Lieberman, Matthew D. Social, Ibid., pp. 145-146.
54 Epley, Nicholas. Mindwise, Ibid., pp. 6-13.
55 Zaki, Jamil & Kevin Ochsner. *You, Me, and My Brain*, Ibid., p. 32.
56 Lieberman, Matthew D. Social, Ibid., p. 127.
57 Epley, Nicholas. Mindwise, Ibid., p. 9.
58 Epley, Nicholas. Mindwise, Ibid., pp. 10-11.

CHAPTER 26

1 Bozarth, Jerold D. *Rogerian Empathy in an Organismic Theory: A Way of Being*, p. 104 citing Ickes (1997). The Social Neuroscience of Empathy, edited by Jean Decety & William Ickes. Bradford BBook, MIT Press, 2011.
2 Eisenberg, Nancy & Natalie D. Eggum. *Empathic Responding: Sympathy and Personal Distress*, Ibid., p. 71 citing Eisenberg et al. (1991). The Social Neuroscience of Empathy, Ibid.

[3] Nickerson, Raymond S., Susan F. Butler & Michael Carlin. *Empathy and Knowledge Projection*, p. 43. The Social Neuroscience of Empathy, Ibid.

[4] Goubert, Liesbet, Kenneth D. Craig & Ann Buysse. *Perceiving Others in Pain: Experimental and Clinical Evidence on the Role of Empathy*, p. 153. The Social Neuroscience of Empathy, Ibid.

[5] Carter, C. Sue, James Harris & Stephen W. Porges. *Neural and Evolutionary Perspectives on Empathy*, p. 169. The Social Neuroscience of Empathy, Ibid.

[6] Batson, Daniel. *These Things Called Empathy: Eight elated but Distinct Phenomena*, pp. 3-8. The Social Neuroscience of Empathy, Ibid.

[7] Hatfield, Elain, Richard L. Rapson & Yen-Chi L. Le. *Emotional Contagion and Empathy*, p. 19 citing Decety & Jackson (2004). The Social Neuroscience of Empathy, Ibid.

[8] Lieberman, Matthew D. Social, pp. 152-157. Crown Publishers, 2013.

[9] Watson, Jeanne C. & Leslie S. Greenberg. *Empathic Resonance: A Neuroscience Perspective*, p. 126 citing Dansinger et a. (2006). The Social Neuroscience of Empathy, Ibid.

[10] Siegel, Daniel, J. M.D. Pocket Guide to Interpersonal Neurobiology, p. Al-29. W.W. Norton & Company, 2012.

[11] Lieberman, Matthew D. Social, Ibid., pp. 154-155.

[12] McCall, Cade & Tania Singer. *Empathy and the Brain*, p. 207. Understanding Other Minds, edited by Simon Baron-Cohen et al. Oxford University Press, 2013.

[13] Cozolino, Louis. The Neuroscience of Human Relationships, p. 203. W.W. Norton & Company, 2006.

[14] Carter, C. Sue, James Harris & Stephen W. Porges. *Neural and Evolutionary Perspectives on Empathy*, Ibid., p. 174.

[15] Gazzaniga, Michael S., Richard B. Ivry & George R. Mangun. Cognitive Neuroscience, 3rd edition, p. 617. W.W. Norton & Company, 2009.

[16] Young, Liane & Adam Waytz. *Mind Attribution is for Morality*, p. 99. Understanding Other Minds, Ibid.

[17] McCall, Cade & Tania Singer. *Empathy and the Brain*, Ibid., p. 202.

[18] Lieberman, Matthew D. Social, Ibid., p. 157.

19 Lieberman, Matthew D. Social, Ibid., pp. 159-160.
20 Zaki, Jamil & Kevin Ochsner. *You, Me, and My Brain*, p. 33. Social Neuroscience, edited by Alexander Todorov et al. Oxford University Press, 2011.
21 Perry, Anat & Simone Shamay-Tsoory. *Understanding Emotional and Cognitive Empathy: A Neuropsychological Perspective*, p. 179. Understanding Other Minds, Ibid.
22 Siegel, Daniel J. M.D. Mindsight, p. 59. Bantam Books, 2010.
23 Perry, Anat & Simone Shamay-Tsoory. *Understanding Emotional and Cognitive Empathy: A Neuropsychological Perspective*, Ibid., p. 179.
24 Perry, Anat & Simone Shamay-Tsoory. *Understanding Emotional and Cognitive Empathy: A Neuropsychological Perspective*, Ibid., 184.
25 Perry, Anat & Simone Shamay-Tsoory. *Understanding Emotional and Cognitive Empathy: A Neuropsychological Perspective*, Ibid., p. 182.
26 Watson, Jeanne C. & Leslie S. Greenberg. *Empathic Resonance: A Neuroscience Perspective*, Ibid., pp. 132-133.
27 Shamay-Tsoory, Simone G. *Empathic Processing, Its Cognitive and Affective Dimensions and Neuroanatomical Basis*, p. 226. The Social Neuroscience of Empathy, Ibid.
28 Watson, Jeanne C. & Leslie S. Greenberg. *Empathic Resonance: A Neuroscience Perspective*, Ibid., p. 128.
29 McCall, Cade &Tania Singer. *Empathy and the Brain*, Ibid., p. 205.
30 McCall, Cade & Tania Singer. *Empathy and the Brain*, Ibid., p. 205.
31 Carter, C. Sue, James Harris & Stephen W. Porges. *Neural and Evolutionary Perspectives on Empathy*, Ibid., p. 178, citing a review by Chakrabarti and Baron-Cohen (2006).
32 Ickes, William. *Empathic Accuracy: Its Links to Clinical, Cognitive, Developmental, Social, and Physiological Psychology*, p. 65 citing Ickes et al. (2000). The Social Neuroscience of Empathy, Ibid.
33 Nickerson, Raymond S., Susan F. Butler & Michael Carlin. *Empathy and Knowledge Projection*, Ibid., p. 49.
34 Watson, Jeanne C. & Leslie S. Greenberg. *Empathic Resonance: A Neuroscience Perspective*, Ibid., p. 131.

35 Watson, Jeanne C. & Leslie S. Greenberg. *Empathic Resonance: A Neuroscience Perspective*, Ibid., p. 132 citing de Vignemont & Singer (2006).
36 Ickes, William. *Empathic Accuracy: Its Links to Clinical, Cognitive, Developmental, Social, and Physiological Psychology*, Ibid., p. 60 citing Barone et al. (2005).
37 Ickes, William. *Empathic Accuracy: Its Links to Clinical, Cognitive, Developmental, Social, and Physiological Psychology*, Ibid., p. 62 citing Gesn & Ickes (1999); Hall & Schmid Mast (2007).
38 Ickes, William. *Empathic Accuracy: Its Links to Clinical, Cognitive, Developmental, Social, and Physiological Psychology*, Ibid., p. 64 citing Thomas, Fletcher, & Lange (1997).
39 Goubert, Liesbet, Kenneth D. Craig & Ann Buysse. *Perceiving Others in Pain: Experimental and Clinical Evidence on the Role of Empathy*, Ibid., p. 155 citing W. Ickes, personal communication (2005).

CHAPTER 27

1 Van der Kolk, Bessel A. M.D. The Body Keeps the Score, p. 206, citing Joseph LeDoux. Viking, 2014.
2 Fuster, Joaquin M. The Prefrontal Cortex, 4th edition, pp. 340-344. Elsevier, 2008.
3 Additional information on wise questions is available through The Public Conversations Project.
4 Van der Kolk, Bessel A. The Body Keeps the Score, Ibid., p. 208.
5 Eysenck, Michael W. & Mark T. Keane. Cognitive Psychology, A Student's Handbook, 6th edition, p. 262 citing Schacter & Addis (2007). Psychology Press, 2010.
6 McVeigh, Brian J. *The Self as Interiorized Social Relations*, p. 211. Reflections on the Dawn of Consciousness, edited by Marcel Kuijsten. Julian Jaynes Society, 2006.
7 Wegner, Daniel M. Who is the Controller of Controlled Processes? p. 30. The New Unconscious, edited by Ran R. Hassin et al. Oxford University Press, 2005.

Index

action potentials, 36, 37-38, 72, 168
adverse childhood experience (ACE) test, 153
amydala
    and fear, 107-08, 109, 123, 131, 133, 136-37, 142-45
    and prefrontal cortex interaction, 62, 63, 76, 109-10, 124-25, 129, 144, 147
    and top-down networks, 109, 118, 136, 144, 206-07
    as a limbic structure, 6, 45, 55
    as a memory system, 134, 176
    fully mature during gestation, 107
    functions and interesting facts, 55
    interconnected nuclei of, 136-37, 146
    low road and high road, 143
    projections to and from, 136-37
    reciprocal activation with cortex, 109, 144
    role in emotion, 126-27, 136-37, 142-45
    role in social cognition, 137, 218, 231
    warning, threat, and danger detection system, 75, 136-37, 145-48
    see also orbitofrontal cortex
anger, 130, 132
    see also brain-systems in the brain
anterior cingulate cortex, 9, 46, 88, 197
    dorsal anterior cingulate cortex, 219, 232-33
    functions and interesting facts, 57
    physical and social pain, 103, 232-33
    role in early organization of self, 110
    role in consciousness, 201
attachment, 114-121

adult attachment interview, 117
and forming relationships, 119-21
and self-control, 117-18
and self-esteem, 118-19
attachment bond, 114
categories of attachment, 115-16
experience of safety, 114-15
long-term consequences of attachment, 116-17
maternal attunement, 75, 106-07
mis-attunement, 116
polyvagal theory, 119-21
schema, 121
see also parenting
autonomic nervous system, 3, 5, 27, 33, 40, 44, 75, 119, 123, 129, 146, 147, 190
axons, see neurons

barely know ourselves, 188, 198
basal ganglia, 7, 46, 48, 130
functions and interesting facts, 55
habitual behavior and, 191-92
role in finding and forming patterns, 84
role in subcortical approach to thinking, 93-95
role in taking action, 87-88
bed nucleus of the stria terminalis, 130

belief perseverance bias, 192
biases, 177, 192-93
binary instinct, 190, 222
bottom-up pathways, 8, 61, 91, 206, 259
brain
and the mind, 28, 64, 89-90, 97, 140-41
argument against a social brain, 220
development of, 103-05
evolution and size of, 14-17
forming causal associations, 74, 85, 91, 111, 165, 204
how networks support shared human functions, 53
how the brain is not like a computer, 72-73
how the brain seeks certainty, 27, 76, 84, 164
initial genetically programmed connectivity in, 101, 103-05
key principles of, 197-98, 213, 254-57
left hemisphere interpreter, 155-57, 223
myth of using only 10% of, 10
pattern finding and making, 83-85
psychology of, 7-8
purpose of, 7, 97-98
self-organizing nature of, 27, 52, 169

small world architecture, 52-53, 68
social nature of, 8, 110, 213-15, 257
special properties of, 28
systems in the brain
    fear system, 26, 133
    panic system, 26, 133
        care circuits, 102
    rage system, 26, 132
    seeking systems, 25-26, 73, 132, 163
three goals of the brain and the mind, 98
triune brain, 12-14
unique human capabilities attributable to, 17, 72-77
variability among individuals, 53, 159, 262-63
brainstem, 43-44, 146
    as fully formed at birth, 107
    medulla, pons, midbrain, 43
    reticular formation, 44
Broca's area, 12, 20, 234

cannot be neutrals, 209
cerebellum, 5, 44, 88, 93-95
    role in procedural learning, 179
cerebral cortex, 5
    computational theory of mind and, 67-68
    description of, 46-49
    development of, 103, 107-08
    functional specialization in, 52
    functions and interesting facts, 58-60
    reciprocal activation with amydala, 109, 144
    role in emotion and emotional processing, 127, 134-36, 143-45, 145-47
    role in memory, see neocortex
    see also neocortex
central nervous system
    description of, 33
    relationship to movement, 3
cingulate cortex, 6, 46, 123, 217
    evolution of resonance behaviors, 102-03
    functions and interesting facts, 57
cognition
    cognitive representations and perceived threat, 24, 135
    computational theory of mind, 65-68, 152
    critical windows and conceptual learning, 110-12
    definition of, 8
    description of, 45, 66
    embodied cognition, 69-70
    outside conscious awareness, 189
    see also knowing; social cognition

communication and language, 233-34, 245
    and the two hemispheres of the brain, 21
conscious awareness, 155, 187, 205
    see also consciousness
consciousness, 201-09
    and change, 207
    and control, 5, 206-07
    and free will, 204-06
    and intelligence, 202-04
    and the prefrontal cortex, 188, 201
    as contrasted with the unconscious, 188-89
    as needing to be engaged, 201, 204
    attention and, 203
    bicameral mind and birth of, 223-24
    description of, 207-08
    easy and hard problem, 89
    mind-body problem, 89
    neural correlates of, 206
    outside traditional western paradigm, 208
    role in emotion, 136
    see also conscious awareness
conservative property of the brain, 28, 63, 72, 77, 85, 169
corpus callosum, 21, 46, 59, 140
cortical maps, 11-12
    role in perception, 67
    see also homunculus

cortico-centrism, 62
cortisol
    and learning, 74, 76
    and threat response, 145-48
creativity, 78, 88, 135
    and the two hemispheres of the brain, 22
    as contrasted with consciousness and intelligence, 203-04
    insight and, 204
critical windows, 110-12
cultural-historical psychology, 233
culture, 221-29
    bicameral mind and birth of consciousness, 223
    group membership and, 225-26
    purpose of a self in, 224-25
    sculpting the mind, 222-27
    role of self-regulation, 226-27

default network, 39, 62, 216
defenses, 135, 191
dendrites, *see* neurons
diencephalon, 44
dopamine, 74, 108, 228
    habitual behavior and prediction-error, 197
dorsolateral prefrontal cortex, 48, 59, 113, 144
    executive functions, 95, 113
    functions and interesting facts, 59

role in supporting working
memory, 7, 48, 61, 88, 92,
93, 134, 176
in the perception-action cycle,
92-93
see also working memory; prefrontal cortex

egocentric bias, 192
embodied cognition, 69-70
embodied property of the brain,
28, 72, 77, 213
emergent property of the brain, 28,
72, 77, 89, 169, 201
emotion, 122-51
amygdala and, 126-27, 136-37,
142-45
amygdala and orbitofrontal cortex, 62, 126, 127, 147
appraisal, 40
as an interpretation, 126
cognitive appraisal, 135
as contrasted with feelings, 40,
78, 125-26
as markers of affective value,
124
as patterns of arousal, 124-25
as the foundation of reason, 123
basic and social emotions, 125
beliefs as sensations in the
body, 124
brain circuitry of, 126
brain regions involved in, 127-
30, 134
cortex and, 134-36
decision-making and, 137-39
definition of, 123
emotional development,
113-21
emotional operating systems,
132-33
emotional re-action, 142, 147
implicit and explicit emotional
learning, 133-34
inhibitory forces in, 109, 118,
144-45, 147-48, 150
physiological changes in the
body, 123, 125-26, 145-48
role of consciousness in, 136
role in thoughts and thinking,
95
storage systems for, 127, 134
types of, 130-31
see also amygdala; cerebral
cortex; fear; stress and the
brain
empathy, 247-53
accuracy of, 252
affect matching, 249
and taking action, 248-49
emotional and cognitive empathy, 250-51
in females and males, 252
multiple definitions of, 248
regions of the brain involved in,
249
enriched environment, 108
evolution

and self-awareness, 154-55
basic building blocks of, 221
emotion and, 123-24
endowing the brain with three special properties, 26-28
history of a larger human brain, 14-17
human universals, 25
instincts, 23-24, 190-91
nurturance and, 102-03
of consciousness, 208
of resonance behaviors, 102-03
organization of the brain, 51-53
origins of human sociality, 221-22
polyvagal theory and, 120
unconsciousness as a gift of, 188
value of amygdala and hippocampus, 45
exaptation, 14, 52, 239
executive control as contrasted with executive functions, 95
explicit memory, 173, 174, 176, 178
compared to implicit, 176-79

fear
amygdala and, 109, 131, 133, 136-37, 142
conditioned fear, 133, 142-45
early learning bias towards fear, 109, 131
fear system, 26, 133
low and high road, 143
threat response, 75-76, 145-48
see also emotion; parenting
fight/flight/freeze, 3
alternatives to, 147-48
and the polyvagal theory, 120
as a biological response, 24
as part of the fear system, 26
threat response and, 145-48
frontal lobe, 5, 6, 47-48
functions and interesting facts, 58
see also prefrontal cortex
functional magnetic resonance imaging, 17-18
functional specialization, 18, 52
fundamental attribution error, 198
future of helping professions, 258-65
begin with the body, 258-60
bring mind patterns into awareness, 260-62
focus on the emergent, 263-64
increase self/other awareness, 262-63
moving beyond mind, 264-65

genes, 29, 101, 103, 104
glial cells, 33, 104

habits and habitual behavior, 195-97
hardwired or inborn human capabilities
   emotional operating systems, 25-26
   human drives and universals 24-25
   instincts, 23-24, 190-91
   major pathways in the brain, 22-23
hemispheres of the brain
   communication and language and, 21
   creativity and, 21
   hemispheric specialization, 20
   learning and, 167, 171
   myth of left and right-brained people, 20-21
   processing faces and, 231
heuristics, 191
hierarchical programming, 64
hippocampus, 6, 45, 107
   functions and interesting facts, 56
   role in emotion, 128, 134
   role in memory creation, 54, 171-72, 176-79
homeostasis, 5, 40
homunculus
   the "homunculus problem", 71
   topographical maps, 11-12, 67
hormones, 44, 74, 126, 146, 148
human tendencies, five overarching, 8-9
hypothalamus, 44, 118, 124, 130, 137
   HPA axis, 44, 75, 109, 146
   launching a physiological response, 143, 145-47
   role in emotion, 128

implicit learning and memories, 121, 123-24, 143, 164, 191
   implicit emotional learning, 133-34
   implicit prejudice and stereotypes, 232
implicit memory, 175, 189
   and procedural memory, 176, 178-79
   beliefs as part of, 179-80
   compared to explicit memory, 176-79
inferences, 85
insight, 78, 195
   neural basis of, 204
insula
   functions and interesting facts, 56
   role in early organization of self, 110
   role in emotion, 127-128
   role in social cognition, 217

intelligence
    general, crystallized, fluid, and metacognition, 202
intuition, 78, 177

knowing, 9, 76-77, 154, 157, 255
    and certainty, 27, 76, 79, 84, 154, 164
    and consciousness, 201
    and intelligence, 203
    as a mental sensation, 76, 79
    beliefs as contrasted with knowledge, 179-80, 182-83
    known truths, 154, 182-83
    sense of self and, 76, 110
    social cognition as, 230
    see also cognition
knowing another indirectly, 159-60

language, *see* communication and language
law of/path of least effort, 27, 28, 170, 181, 189, 207, 256
learning
    belief-driven and knowledge-driven approaches to learning, 179-80
    early learning bias towards fear, 109, 131
    goal of, 163, 256
    Hebbian learning, 53
    how the brain learns, 165-68
    learning at the cellular level, 168-69
    learning fairness, 112
    practice, rehearsal, and elaborate rehearsal, 166-67
    role of working memory in, 163, 165-66
    sensory registry, 165, 173
    transfer, 165
    unlearning, 145, 170
    see also neuroplasticity
left hemisphere interpreter, 155-57, 223
limbic system, 6
    description of, 45-46, 123
    development of, 107
    functions and component regions, 54-56
    not a system, 124
lobes of the brain, 47
long-term memory, *see* learning; memory
long-term potentiation, 49, 76, 149, 168

major pathways in the brain
    fear and stress pathway, 22
    pain pathway, 22
    regulatory pathways, 23
    reward pathways, 23
meaning
    and the brain, 73-75

how learning creates, 164
role of others in determining, 224-26
medial prefrontal cortex
   and inner body awareness, 259
   and self-reflection, 154-55, 218
   see also orbitofrontal cortex
memory, 171-183
   autobiographical, 178, 181
   biological systems comprising, 171
   confabulation, 181
   consolidation, 172
   encoding, 171
   episodic, 6, 153, 174-75, 176, 182
   explicit as contrasted with implicit, 176-79
   failure of, 180-81
   goal of, 163, 171
   long-term, 6, 163, 167, 173-75, 176-80
   multiple trace theory, 182
   procedural, 175, 176, 178-79
   retrieval, recognition, and recall, 163, 172, 174
   semantic, 174, 178
   sense of self and, 182-83, 204
   sensory memory, 165, 173
   storage of, 172, 174
   types of, 173-75
mentalizing, see theory of mind
mental states, see theory of mind
mesocortex, 6
metacognition, 202
mind
   and the brain, 28, 64, 89-90, 97, 140-41
   capable of being observed, 227
   "homunculus problem," 71
   mind patterns, 39, 53, 84, 154, 191, 203, 260
   primitive mind, 208
   role of culture in sculpting, 222-27
   where the self lives, 98, 208-09
   see also brain
mind/body problem, 89
mindfulness, 207
mirror neurons, 74, 215
   mirror system, 242-44, 250-51
models of the brain
   computational theory of mind, 65-68
      and consciousness, 89
      and representational thinking, 85-88, 241, 243
   convergence and synchronization, 69
   hierarchical processing, 64
   modularity, 64
   network dynamics model, 68-70
   nodes, modules, hubs, 68
   parallel distributive networks, 65

recurrent, recursive, and reentrant circuits, 69, 70, 88, 140, 201, 206
modularity, 64
motor neurons, 5, 7, 34, 35, 43, 67
  mirror neurons as, 218
motor system
  and taking action, 87-88
  in the perception-action cycle, 91-93
myelination, 33, 46, 120

neocortex
  description of, 47-48
  premotor cortex, 5, 6, 47, 87-88, 93, 179
  primary motor cortex, 5, 6, 47, 68, 87-88, 93
  projections to the amygdala, 139
  relative size of and preferred group size, 220-21
  role in memory, 171
  somatosensory cortex, 5, 11, 35, 47, 66-68, 128, 249
  supplementary motor cortex, 5, 6, 47, 88
  see also cerebral cortex
neural correlates or signature, 17, 18, 68, 89, 206, 218-19
neural organization
  during development, 103-05
  functional homeostasis, 53
  initial genetically programmed connectivity, 101, 103-05
  see also neuroplasticity
    – experience-based
    role of competition for synapses, 51
    rooted in survival and driven by needs, 51
  self-organizing nature, 27, 52, 169
  small world architecture/connectivity, 52-53, 68
neuro-centrism, 62
neuroception of safety, 120, 259
neurogenesis, 34, 49, 104, 169
  see also neuroplasticity
neurons, 3-4, 34
  axons, 12, 35, 36, 37, 46, 47, 105, 168
  channels, charges, thresholds, action potentials and spikes, 36-37
  computation neurons, 34
  dendrites, 12, 34, 46, 108, 168
  description of, 3-4
  factors that make neurons special, 34
  formation of pathways of communication, 35-39, 254
    role of parenting in, 108
  forms of, 35
  glial cells, 33, 104
    astrocytes, 33

motor neurons, 5, 7, 34, 35, 43, 67
myelination, 33, 46, 120
neuron doctrine, 12
neurotransmitters, 12, 34, 38, 53, 74, 168
sensory neurons, 4, 5, 34, 35
neuroplasticity, 27, 49-50
and stress, 76
as it relates to learning, 163-70
description of, 49
during fetal development, 105
experience-based, 49, 63, 101, 105, 107-11, 114-17, 144
see also past
extent of, 169-70
NMDA receptors, 168
synaptic plasticity, 33, 49, 104-05, 111, 168-69,
neuro-reductionism, 62, 64, 70
neuroscience
definition of, 3, 64
history of, 11-12
modern-day research in, 17-18
neurotransmitters, 12, 34, 38, 53, 74, 168

occipital cortex, 47, 49
functions and interesting facts, 59
representational thinking and the vision system, 86-87

olfactory system, 13, 49, 51, 66, 130, 137
orbitofrontal cortex, 7, 45, 48, 88, 92, 93, 124, 197, 216, 249
functions and interesting facts, 60
parenting and the development of self-control, 109, 113, 117-18
role in emotion, 62, 126, 127, 134, 137-38, 147, 150
role in decision-making, 137-38
role in integrating emotions and memories into thoughts, 45, 62
see also emotion; amygdala; prefrontal cortex; self-regulation/control
oxytocin and vasopressin, 45, 108, 228

parallel distributive networks, 65
parasympathetic nervous system, 3, 75, 120, 127, 146, 150, 249
parenting, 101-21
attachment relationships, 114-17
critical windows, 110-12
early learning bias towards fear, 109, 131
effects of sustained shaming, 119

emotional learning and the experience of safety
attachment and, 114-16
response to baby's distress, 109, 117-18
safety danger-continuum, 101, 109, 131, 142
enriched environment, 108
evolution of nurturance, 106-08
first sense of knowing, 110
first sense of self, 110
good enough parenting, 112
how parenting builds biology, 108-10
internalized mother, 109
panic system and care circuits, 26, 102, 133
pregnancy and brain development, 103-05
self-control and, 117-18
self-esteem and, 118-19
social engagement and, 107-08, 110, 119-21
see also attachment; polyvagal theory
parietal cortex, 47, 48, 110
functions and interesting facts, 58
role in consciousness, 201
role in vision, 86
past, 152-160
and the amygdala, 142-43, 144
as attachment schema, 121
filter of past, 157-59, 189
importance of early experience, 106-12, 152-53
influences of, 17, 73, 78, 134, 140-41, 158-59, 176, 189-90, 200, 258
learning and past, 164, 165, 168
memory and past, 171, 173, 175, 178
past experience
role in perception-action cycle of thinking, 91-93
role in representational thinking, 85-87
role in self-regulation, 144
patterns and past, 83-84
social cognition and, 213, 239, 242, 243
see also neuroplasticity-experience-based
patterns
amygdala and finding patterns among people, 137
attachment relationships as a pattern, 114
culture as patterns, 222
emotions as patterns of arousal, 124-25
habits as patterns, 195-97
social cognition as built on, 213

thinking as based on, 83-85, 213, 255
perception
    as a computational process, 65-68, 86-88, 152
    as constructed, 39-41, 86-87, 140-41, 158-59, 189-93, 258-59
    in the perception-action cycle, 92
    types of sensory receptors, 66
    see also reality; past
perception-action cycle
    five principle cognitive functions, 92
    perceptual memory and executive memory, 92-93
    unique role for the prefrontal cortex, 92
peripheral nervous system, 3, 33, 43
phonemes, 233
phrenology, 11
plasticity, see neuroplasticity; synaptic plasticity
polyvagal theory, 119-20
prefrontal cortex, 6, 13, 16, 48, 59-60
    and amygdala interaction, 62, 63, 76, 109-10, 124-25, 129, 144, 147
    and healing, 75-76
    and intelligence/attention, 203
    and self-recognition, 154, 218
    and stress, 76
    and the representational approach to thinking, 85-88
    and the perception-action approach to thinking, 91-93
    as intentional more than rational, 48, 62-63, 95, 260
    dorsolateral prefrontal cortex and orbitofrontal cortex, 48, 62, 95, 113
    extensive projections to and from, 60-61
    functions and interesting facts, 59
    generative powers, 260-62
    in greater detail, 60-63
    limitations of, 62-63, 96, 134-36
    orbitofrontal and orbitomedial prefrontal cortex, 45, 60, 109, 113, 118, 127, 147, 216, 249
    role in integrating thoughts and emotions, 62
    role in self-regulation/control, 76, 96, 109, 118, 144, 206-07
    see also orbitofrontal cortex; dorsolateral prefrontal cortex; recurrent, recursive, and reentrant circuits
premotor cortex, see neocortex

prepared learning, 24
primary motor cortex, see neocortex
priming, 192
pruning, see synaptic realignment

reality
    creating our experience of, 7-8, 112
    experience as doubly removed from, 159
    how real is, 140-41
    see also perception
reciprocal altruism, 221
    and human nature, 228-29
recurrent, recursive, and reentrant circuits, 69, 70, 88, 140, 201, 206
remembering, see memory
representational thinking, 85-88, 152
    and taking action, 87-88
    and theory of mind, 242-44
    and the vision system, 86-87

Sally/Anne false belief task, 241
second-order change, 207
self-regulation/control
    as serving a social purpose, 226-27
    attachment and, 117-18
    consciousness and, 206-07
    inhibition, 109, 118, 144-45, 147-48, 150
    part of social tool kit, 239
    prefrontal cortex and, 76, 96, 109, 118, 144, 206-07
    role of parenting in the development of, 109
    top-down regulatory networks, 109, 118, 136, 144, 206-07
    downside of, 207
    unconscious, 194
    see also top-down pathways
sense of self, 71
    belonging, 225, 228-29
    everything as a projection of, 155
    evolution and self-awareness, 155
    experience of agency, 41, 71, 97, 154, 204-05
    first sense of self, 110
    free will, 204-06
    knowing and, 76, 110
    maintaining a coherent sense of, 41, 98, 156, 182-83
    memory and, 182-83
    neural correlates related to, 218-19
    self-esteem, 118-19
    sense of "I", 153
    social role of, 224-25
    story of "me," 41, 153-55, 182, 224
        and the left hemisphere interpreter, 156

unconscious assumptions about living, 197-98
sensory neurons, 4, 5, 34, 35
septal area, 128, 217, 250
simulation, 215
   and empathy, 249, 250-51
   and theory of mind, 242-44
"sins of memory", 180-81
social cognition, 213-35
   a brain that's also social, 220
   accuracy of, 213-15, 245, 252
   areas of the brain that participate in, 215, 216-18
   default system and, 216
   definition and description, 213-14
   early learning and social engagement, 111-12
   emotional and cognitive empathy, 250-51
   knowing as a part of, 76
   making sense of another human being, see theory of mind
   otherness, 84, 112, 159
   processing faces, 230-31
   processing language and communication, 233-34
   processing rules, deception, disgust, aggression, and revenge, 234-35
   processing social categories, 231-32
   reading mental states of others, see theory of mind
   role of group membership, 225-26
   role of mirror neurons and spindle cells, 74, 103, 215, 242-44
   social and non-social thinking, 216, 239
   social pain, 219, 232-33
   social tasks and associated neural correlates, 218-19
   see also theory of mind; empathy
social neuroscience
   brain as innately social, 24, 101, 110, 215, 257
   primary neural regions, 215
   research findings, 216
   see also social cognition
social relationships
   and social conformity, 224-27
   impact of unconscious influences, 200
   see also social cognition
somatic marker hypothesis, 138
somatosensory cortex, see neocortex
spinal cord, 3, 4, 5, 7, 33, 43, 48
   and perception, 66-68
   and reflexes, 87
spindle cells, 74, 103, 219
stereotyped expression, 125

stereotypes, 191-92
stress and the brain
    and the prefrontal cortex, 76
    as a re-action, 148-49
    contrasted with a threat response, 150
    effects of, 75-76, 148-49
    role of parenting and safety, 109
    stress and learning, 76
    types of stressors, 148
subjective inner state, 39-41
    effect of stress on, 148
    relationship to sense of self, 40-41, 204
    viscera, 39, 73, 75, 123, 124
supplementary motor cortex, see neocortex
survival
    as first and foremost, 8
    drive to defend, 24, 135
    social survival, 8
sympathetic nervous system, 5, 75, 120, 127, 146, 148, 150, 249
synaptic plasticity, 33
    and synaptogenesis, 49, 111
    and unlearning, 145
    synaptic realignment and pruning, 103-05
        as it relates to learning, 164, 168-69
        experience-based reorganization, 49, 63, 101, 105, 107-11, 114-17, 144

temporal cortex, 47
    functions and interesting facts, 59
    role in memory, 171, 174
    role in vision, 86
thalamus, 5, 44, 47, 60, 67, 86, 137, 143
theory of apparent mental causation, 204
theory of mind, 215, 237-46
    and interpretation, 245
    as an art, 245
    contrasted with perspective-taking, 239
    definition and description, 237-38
    developing a theory of mind, 239-41
        keys to successful mentalizing, 239
    mentalizing about self and others and related neural correlates, 218
    regions of the brain involved in, 238
    representations and simulation in empathy, 250-51
    representations, 239-41, 242, 243-44
    mirror system, 242-43

two distinct systems, 244
  see also empathy
thinking and the brain
  general, 95-96
  perception-action cycle, 91-93
  representational thinking, 85-88
  social and non-social thinking, 216, 239
  subcortical approach to, 93-94
threat and danger detection system, see amygdala
threat response, 75-76, 145-48
  mitigating factors, 147
  three circuit model for mammals, 120
  two circuit model and children, 115-16
top-down pathways, 8, 61, 91
  regulatory networks, 109, 118, 136, 144, 206-07
  see also self-regulation/control
triune brain
  model of, 12
  problems with, 13-14

unconscious, 187-200
  as contrasted with consciousness, 188
  as preserving what is known, 189
  assumptions about living, 197-98
  automatic body movements, 190, 199-200
  biases, 192-93, 231-32
  complex social behavior, 193
  constructed perception, 190
  deception and, 235
  defenses and heuristics, 191
  filter of past, 190
  implicit re-actions, 191, 199-200
  instincts, 190-91, 199-200
    binary instinct, 190, 222
  new unconscious, 193-95
  priming and stereotypes, 191-92
  role of parenting in creation of, 109
  social cognition skills as, 214

vagus, smart, 119-21
viscera, 39, 73, 75, 123, 124
vision system, 86-87
  as constructive, 86-87

Wernicki's area, 12, 20, 234
wisdom of the brain, 8, 19, 28, 40, 50, 54, 65, 77, 85, 98, 101, 111, 121, 125, 141, 151, 159, 170, 179, 197, 206, 213, 227, 230, 245, 253
working memory
  as a distributive process, 176
  chunking, 175

phonological loop and visual sketchpad, 173, 175
role in goal hierarchies, 88, 93
role of lateral prefrontal cortex, 7, 48, 61, 88, 92, 93, 134, 176
role in learning, 163, 166-67
role in remembering, 171, 172, 175-76

www.ingramcontent.com/pod-product-compliance
Lightning Source LLC
Chambersburg PA
CBHW070222190526
45169CB00001B/42